THE MEAS

STAR

Review of the U.S. Environmental Protection Agency's Science to Achieve Results (STAR) Research Grants Program

Committee to Review EPA's Research Grants Program

Board on Environmental Studies and Toxicology

Division on Earth and Life Studies

NATIONAL RESEARCH COUNCIL
OF THE NATIONAL ACADEMIES

THE NATIONAL ACADEMIES PRESS
Washington, D.C.
www.nap.edu

THE NATIONAL ACADEMIES PRESS 500 Fifth Street, NW Washington, DC 20001

NOTICE: The project that is the subject of this report was approved by the Governing Board of the National Research Council, whose members are drawn from the councils of the National Academy of Sciences, the National Academy of Engineering, and the Institute of Medicine. The members of the committee responsible for the report were chosen for their special competences and with regard for appropriate balance.

This project was supported by Contract 68-C-01-119 between the National Academy of Sciences and the U.S. Environmental Protection Agency. Any opinions, findings, conclusions, or recommendations expressed in this publication are those of the authors and do not necessarily reflect the view of the organizations or agencies that provided support for this project.

International Standard Book Number 0-309-08938-7 (Book)
International Standard Book Number 0-309-51674-9 (PDF)

Additional copies of this report are available from:

The National Academies Press
500 Fifth Street, NW
Box 285
Washington, DC 20055

800-624-6242
202-334-3313 (in the Washington metropolitan area)
http://www.nap.edu

Printed in the United States of America

THE NATIONAL ACADEMIES

Advisers to the Nation on Science, Engineering, and Medicine

The **National Academy of Sciences** is a private, nonprofit, self-perpetuating society of distinguished scholars engaged in scientific and engineering research, dedicated to the furtherance of science and technology and to their use for the general welfare. Upon the authority of the charter granted to it by the Congress in 1863, the Academy has a mandate that requires it to advise the federal government on scientific and technical matters. Dr. Bruce M. Alberts is president of the National Academy of Sciences.

The **National Academy of Engineering** was established in 1964, under the charter of the National Academy of Sciences, as a parallel organization of outstanding engineers. It is autonomous in its administration and in the selection of its members, sharing with the National Academy of Sciences the responsibility for advising the federal government. The National Academy of Engineering also sponsors engineering programs aimed at meeting national needs, encourages education and research, and recognizes the superior achievements of engineers. Dr. Wm. A. Wulf is president of the National Academy of Engineering.

The **Institute of Medicine** was established in 1970 by the National Academy of Sciences to secure the services of eminent members of appropriate professions in the examination of policy matters pertaining to the health of the public. The Institute acts under the responsibility given to the National Academy of Sciences by its congressional charter to be an adviser to the federal government and, upon its own initiative, to identify issues of medical care, research, and education. Dr. Harvey V. Fineberg is president of the Institute of Medicine.

The **National Research Council** was organized by the National Academy of Sciences in 1916 to associate the broad community of science and technology with the Academy's purposes of furthering knowledge and advising the federal government. Functioning in accordance with general policies determined by the Academy, the Council has become the principal operating agency of both the National Academy of Sciences and the National Academy of Engineering in providing services to the government, the public, and the scientific and engineering communities. The Council is administered jointly by both Academies and the Institute of Medicine. Dr. Bruce M. Alberts and Dr. Wm. A. Wulf are chair and vice chair, respectively, of the National Research Council

www.national-academies.org

COMMITTEE TO REVIEW EPA'S RESEARCH GRANTS PROGRAM

Members

HAROLD A. MOONEY (*Chair*), Stanford University, Stanford, CA
RAYMOND C. LOEHR *(Vice Chair),* University of Texas, Austin
ANDERS W. ANDREN, University of Wisconsin, Madison
EDWIN H. CLARK, II, Clean Sites, Inc., Washington, DC
COSTEL D. DENSON, University of Delaware, Newark
JOHN C. ELSTON, Spring Lake Heights, NJ, retired from New Jersey State Department of Environmental Protection
CAROL J. HENRY, American Chemistry Council, Arlington, VA
MARTHA A. KREBS, Science Strategies, Los Angeles, CA
RICHARD F. LEE, Skidaway Institute of Oceanography, Savannah, GA
GERALD VAN BELLE, University of Washington, Seattle
TERRY F. YOUNG, Environmental Defense, Oakland, CA
LAUREN A. ZEISE, California Environmental Protection Agency, Oakland

Staff

EILEEN N. ABT, Project Director
ROBERTA M. WEDGE, Program Director for Risk Analysis
NORMAN GROSSBLATT, Senior Editor
KELLY CLARK, Assistant Editor
JENNIFER E. SAUNDERS, Research Assistant
MIRSADA KARALIC-LONCAREVIC, Research Assistant
LUCY V. FUSCO, Senior Project Assistant

Sponsor

U.S. ENVIRONMENTAL PROTECTION AGENCY

Board on Environmental Studies and Toxicology[1]

Members

Gordon Orians *(Chair)*, University of Washington, Seattle
John Doull *(Vice Chair)*, University of Kansas Medical Center, Kansas City
David Allen, University of Texas, Austin
Thomas Burke, Johns Hopkins University, Baltimore, MD
Judith C. Chow, Desert Research Institute, Reno, NV
Christopher B. Field, Carnegie Institute of Washington, Stanford, CA
William H. Glaze, Oregon Health and Science University, Beaverton
Sherri W. Goodman, Center for Naval Analyses, Alexandria, VA
Daniel S. Greenbaum, Health Effects Institute, Cambridge, MA
Rogene Henderson, Lovelace Respiratory Research Institute, Albuquerque, NM
Carol Henry, American Chemistry Council, Arlington, VA
Robert Huggett, Michigan State University, East Lansing
Barry L. Johnson Emory University, Atlanta, GA
James H. Johnson, Howard University, Washington, DC
James A. MacMahon, Utah State University, Logan
Patrick V. O'Brien, Chevron Research and Technology, Richmond, CA
Dorothy E. Patton, International Life Sciences Institute, Washington, DC
Ann Powers, Pace University School of Law, White Plains, NY
Louise M. Ryan, Harvard University, Boston, MA
Jonathan M. Samet, Johns Hopkins University, Baltimore, MD
Kirk Smith, University of California, Berkeley
Lisa Speer, Natural Resources Defense Council, New York, NY
G. David Tilman, University of Minnesota, St. Paul
Chris G. Whipple, Environ Incorporated, Emeryville, CA
Lauren A. Zeise, California Environmental Protection Agency, Oakland

Senior Staff

James J. Reisa, Director
David J. Policansky, Associate Director
Raymond A. Wassel, Senior Program Director for Environmental Sciences and Engineering
Kulbir Bakshi, Program Director for the Committee on Toxicology
Roberta M. Wedge, Program Director for Risk Analysis
K. John Holmes, Senior Staff Officer
Susan N.J. Martel, Senior Staff Officer
Suzanne Van Drunick, Senior Staff Officer
Eileen N. Abt, Senior Staff Officer
Ellen K. Mantus, Senior Staff Officer
Ruth E. Crossgrove, Managing Editor

---

[1]This study was planned, overseen, and supported by the Board on Environmental Studies and Toxicology.

Other Reports of the Board on Environmental Studies and Toxicology

Cumulative Environmental Effects of Alaska North Slope Oil and Gas Development (2003)
Estimating the Public Health Benefits of Proposed Air Pollution Regulations (2002)
Biosolids Applied to Land: Advancing Standards and Practices (2002)
Ecological Dynamics on Yellowstone's Northern Range (2002)
The Airliner Cabin Environment and Health of Passengers and Crew (2002)
Arsenic in Drinking Water: 2001 Update (2001)
Evaluating Vehicle Emissions Inspection and Maintenance Programs (2001)
Compensating for Wetland Losses Under the Clean Water Act (2001)
A Risk-Management Strategy for PCB-Contaminated Sediments (2001)
Acute Exposure Guideline Levels for Selected Airborne Chemicals (3 volumes, 2000-2003)
Toxicological Effects of Methylmercury (2000)
Strengthening Science at the U.S. Environmental Protection Agency (2000)
Scientific Frontiers in Developmental Toxicology and Risk Assessment (2000)
Ecological Indicators for the Nation (2000)
Modeling Mobile-Source Emissions (2000)
Waste Incineration and Public Health (1999)
Hormonally Active Agents in the Environment (1999)
Research Priorities for Airborne Particulate Matter (4 volumes, 1998-2003)
Ozone-Forming Potential of Reformulated Gasoline (1999)
Arsenic in Drinking Water (1999)
The National Research Council's Committee on Toxicology: The First 50 Years (1997)
Carcinogens and Anticarcinogens in the Human Diet (1996)
Upstream: Salmon and Society in the Pacific Northwest (1996)
Science and the Endangered Species Act (1995)
Wetlands: Characteristics and Boundaries (1995)
Biologic Markers (5 volumes, 1989-1995)
Review of EPA's Environmental Monitoring and Assessment Program (3 volumes, 1994-1995)
Science and Judgment in Risk Assessment (1994)
Pesticides in the Diets of Infants and Children (1993)
Dolphins and the Tuna Industry (1992)
Science and the National Parks (1992)
Human Exposure Assessment for Airborne Pollutants (1991)
Rethinking the Ozone Problem in Urban and Regional Air Pollution (1991)
Decline of the Sea Turtles (1990)

Copies of these reports may be ordered from The National Academies Press
(800) 624-6242 or (202) 334-3313
www.nap.edu

Preface

The Environmental Protection Agency (EPA) is a mission agency established in 1970 to protect human health and to safeguard the natural environment. EPA's regulatory and decision-making role requires that the agency have access to the best available science that is relevant to its mission. In an effort to improve the scientific foundation of its decision-making process, the agency established the Science To Achieve Results (STAR) research grants program in 1995.

The STAR program is a competitive, peer-reviewed, extramural research grants program created to encourage interagency collaboration and to increase EPA's access to the nation's best scientists and engineers in academic and nonprofit research institutions. The program supports research in a variety of fields relevant to EPA's mission, ranging from human health protection to environmental preservation. It is designed to maximize the independence of the researchers supported and to provide an equal opportunity for all researchers to qualify for support.

In 2000, EPA asked the National Research Council to conduct an independent assessment of the STAR program. In response, the Research Council established the Committee to Review EPA's Research Grants Program. In this report, the committee analyzes information provided by EPA, STAR grant recipients and fellows, and other sources to assess the program's scientific merit, effect on the agency's policies and decisions, and overall relevance to EPA's mission. In addition, the committee compares some of the procedural aspects of the STAR program with those of basic and applied research grant programs of other agencies. Finally, the committee recommends ways to enhance the program and improve data collection for future program evaluations.

This report has been reviewed in draft form by persons chosen for their diverse perspectives and technical expertise in accordance with procedures approved by the Research Council's Report Review Committee. The purpose of this independent review is to provide candid and critical comments that will assist the institution in making its published report as sound as possible and to ensure that the report meets institutional standards of objectivity, evidence, and responsiveness to the study charge. The review comments and draft manuscript remain confidential to protect the integrity of the deliberative process. We wish to thank the following people for their review of this report: William Glaze, University of North Carolina, Chapel Hill, North Carolina; Bernard D. Goldstein, University of Pittsburgh, Pittsburgh, Pennsylvania; Mark A. Harwell, Florida A&M University, Tallahassee, Florida; George Lucier (retired), National Institute of Environmental Health Sciences, Pittsboro, North Carolina; Perry L. McCarty, Stanford University, Stanford, California; Paul G. Risser, The Oklahoma State System of Higher Education, Oklahoma City, Oklahoma; Joan B. Rose, Michigan State University, East Lansing, Michigan; Jane Warren, Health Effects Institute, Boston, Massachusetts.

Although the reviewers listed above have provided many constructive comments and suggestions, they were not asked to endorse the conclusions or recommendations, nor did they see the final draft of the report before its release. The review of this report was overseen by Bailus Walker, Jr., Howard University Medical Center, Washington, D.C. Appointed by the Research Council, he was responsible for making certain that an independent examination of the report was carried out in accordance with institutional procedures and that all review comments were carefully considered. Responsibility for the final content of the report rests entirely with the committee and the institution.

The committee gratefully acknowledges the following people for making presentations to it: Peter Preuss, Paul Gilman, Timothy Oppelt, John Bachmann, Patricia Bradley, Elizabeth Bryan, William Farland, and James Hanlon, EPA; Robert Huggett, Michigan State University; John Wanska, General Accounting Office; Jerry Elwood, Department of Energy; Peter Johnson, U.S. Department of Agriculture; Claudia Thompson, National Institute of Environmental Health Sciences; Susan Cozzens, Ann Bostrom, and Alan Porter, Georgia Institute of Technology; Penny Firth, National Science Foundation; and Nils Newman, IISCO. The committee also wishes to thank the following, who provided further background information:

Theodore Just, Jack Puzak, Jeffrey Harris, Manju Gupta, Terry Simpson, Matthew Clark, Gina Perovich, Shirley Hamilton, and James Gentry, EPA; Claudia Magdalena Abendroth, Office of Management and Budget; Jeanne Powell, National Institute of Standards and Technology; Julia Melkers, University of Chicago; Christopher Allen, University of Vermont; Robert Selden, Air Force Scientific Advisory Board; Leslie Peasant, Air Force Office of Scientific Research; James Coleman, Nevada National Science Foundation; and Deborah Stine and Scott Weidman, National Research Council. In addition, the committee gives special thanks to the EPA project officers in three research programs—Endocrine Disruptors, Elaine Francis and David Reese; Particulate Matter, Gail Robarge and Stacey Katz; and Ecological Indicators, Barbara Levinson—who were available to discuss their programs in detail. We are also grateful to the many National Center for Environmental Research (NCER) staff who invested extensive time and effort in responding to all the committee's requests for information. Finally, we appreciate the contributions of the numerous EPA STAR grantees and fellowship recipients who provided input on the program.

The committee is grateful for the assistance of the Research Council staff in preparing this report: Eileen Abt, project director; James Reisa, director of the Board on Environmental Studies and Toxicology; Roberta Wedge, program director for risk analysis; Jennifer Saunders and Mirsada Karalic-Loncarevic, research assistants; Ruth E. Crossgrove, managing editor; Norman Grossblatt, senior editor; Kelly Clark, assistant editor; Lucy Fusco and Bryan Shipley, senior project assistants; and Robert Policelli and Tamara Dawson, project assistants.

Finally, I thank the members of the committee for their dedicated efforts throughout the development of this report.

Harold Mooney
Chair, Committee to Review EPA's
Research Grants Program

Contents

The Measure of STAR

Review of the U.S. Environmental Protection Agency's
Science to Achieve Results (STAR)
Research Grants Program

Summary

In an effort to improve the scientific foundation of the Environmental Protection Agency (EPA) decision-making process, EPA's Office of Research and Development (ORD) created the Science To Achieve Results (STAR) program in 1995. The STAR program is a competitive, peer-reviewed, extramural research grants program intended to increase the agency's access to the nation's best scientists and engineers in academic and other nonprofit research institutions. It supports research pertaining to human health and the environment and is designed to maximize the independence of the researchers supported and to provide an equal opportunity for all researchers to qualify for support.

The STAR program, currently managed by ORD's National Center for Environmental Research (NCER), is integrated into EPA's overall research program through extensive planning and consultation with the agency's other research centers and laboratories and its program and regional offices. The research sponsored by the STAR program allows the agency to fill information gaps that are not addressed completely by its intramural research program and to respond to new issues that the EPA laboratories are not able to address.

The research support awarded by the STAR program is of three main kinds: grants awarded to individual investigators or small groups of investigators, grants awarded to multidisciplinary (and sometimes multi-institutional) research centers, and fellowships to support graduate work (at the master's and Ph.D. levels) in environmental sciences. The program has been funded at about $100 million per year over the last few years and accounts for 15-20% of ORD's research budget. The program has leveraged its funds by forming partnerships with other agencies that support similar kinds of research. Since it was established, the components and management of the program have adapted in response to changing agency needs, experience gained in operating the program, and external reviews.

CHARGE TO THE COMMITTEE

In 2000, EPA asked the National Research Council to conduct an independent assessment of the STAR program. In response, the Research Council established the Committee to Review EPA's Research Grants Program, which prepared this report. The committee was given the following task:

> The NRC committee will conduct a program review of EPA's Science To Achieve Results (STAR) competitive extramural research grants program. Using information to be obtained from EPA, STAR grant recipients, and other sources, the committee will assess the program's scientific merit, its demonstrated or potential impact on the agency's policies and decisions, and other program benefits that are relevant to EPA's mission. The committee will recommend ways to enhance the program's scientific merit, impact, and other benefits. In the context of other relevant research conducted or funded by EPA, and in comparison with other basic and applied research grant programs, this assessment will address the STAR program's research priorities, research solicitations, peer-review process, ongoing research projects, results and dissemination of completed research, and other aspects to be identified by the committee.

In undertaking its review, the committee held three public sessions in which it heard presentations about the STAR program by EPA officials and others. The EPA officials represented NCER and other EPA research and program offices. The public sessions included presentations by representatives of other federal agencies that support extramural research and by experts in evaluating research programs. Committee members also interviewed STAR project officers and STAR grant and fellowship recipients and attended STAR sponsored workshops and meetings. NCER staff provided the committee with substantial amounts of information regarding the operation and financing of the program.

THE COMMITTEE'S EVALUATION

The committee's evaluation of the STAR program focused on the program's quality, relevance, and performance as described in the recent Office of Management and Budget (OMB) guidelines on evaluating research programs. The committee used metrics that grew out of its review of information available from EPA and of metrics used by EPA and other organizations. The metrics, which are both quantitative and qualitative, assisted the committee in forming judgments regarding the scientific merit of the program and its impact on the agency.

The committee recognizes that the STAR program is still too young to provide all the information needed for a full evaluation of the extent, impact, and value of its activities. Evaluation of research results is difficult and requires substantial elapsed time; for a given topic, it can take 3-5 years from the initiation of laboratory or field experiments to the analysis and publication of results. Considerably more time must elapse to realize the impact of published research on the scientific and regulatory communities. Nevertheless, the committee judged that it had sufficient information to evaluate how the STAR program operates and its value to the nation's overall environmental research and management efforts.

To effectively communicate its findings in this summary, the committee developed and addressed a series of specific questions that it believed would be of greatest interest to the audience of this report. On the basis of its evaluation, the committee unanimously arrived at the following conclusions and recommendations.

Should the STAR program continue to be part of the ORD research program?

Finding. EPA requires a strong and balanced science and technology research program to fulfill its mission properly. The STAR program is an important part of the overall EPA research program.

Several previous reports by EPA and the National Academies have addressed the question of whether EPA should have its own research program or rely on research results developed elsewhere. Those reports all concluded emphatically that EPA needs its own strong research program to meet the needs of its mission. The committee concurs with that conclusion.

The STAR program is EPA's preeminent program that solicits independent scientific and technologic research from the nation's best academic and other nonprofit research institutions. The program has established and maintains a high degree of scientific excellence. By funding the majority of its research efforts through broadly advertised, competitive grants, the STAR program provides the agency access to independent information, analyses, and perspectives.

The research portfolio of the STAR program is derived directly from the strategic plans of EPA and ORD and from ORD's more-detailed research strategies that address particular topics. It is an integrated part of EPA's research program. The STAR program provides the agency access to a broad community of researchers, allows it to fund research at the cutting edge of science, and assists it in addressing information gaps that it does not have the internal resources to address properly. The STAR program also encourages its grantees to disseminate their research results widely to promote their rapid and widespread use.

For all those reasons, STAR research effectively expands the nation's scientific foundation for protecting human health and the environment. Moreover, by expanding environmental research and analysis capabilities in many of the nation's academic and other nonprofit research institutions and by supporting young scientists interested in environmental research, the STAR program actively expands the nation's environmental-science infrastructure.

Recommendation. The STAR program should continue to be an important part of EPA's research program.

What is the unique contribution of the STAR program?

Finding. The STAR program funds important research that is not conducted or funded by other agencies. The STAR program has also made commendable efforts to leverage funds through establishment of research partnerships with other agencies and organizations.

The STAR program provides EPA with access to independent research that is directly relevant to its mission. The program makes strong efforts to ensure that the results of its research are expeditiously communicated to relevant EPA program offices and to other potential users. The STAR program gives primary potential users of research results a unique role in helping to plan the research and to identify the specific high-quality proposals that will be of greatest value to them. The exploratory and core research that the program sponsors alerts the agency to possible emerging issues, providing more opportunity for the agency to consider how it might best address them.

Much of the research funded by STAR would not have been undertaken without the program, because it is not conducted or funded by other agencies. For instance, EPA is one of the few agencies that provide extramural funding for examining the impacts of endocrine disruptors on ecosystem processes. The STAR ecologic-indicators program is the primary source of support of research on the development of water-quality indicators for biologic monitoring. The interdisciplinary centers that STAR has supported also represent an innovative approach to supporting research that will be most relevant for environmental decision making in several important topics.

Finally, the STAR program has been successful in working with other agencies that have similar or complementary research interests through research partnerships and in obtaining supplementary funding. That not only leverages additional funds for research projects of interest to STAR but also helps to increase the partner agencies' awareness of the pertinent issues and information needs of EPA. The STAR program's ability to establish partnerships has increased as more funds have been allocated to it.

Recommendation. STAR should continue to partner with other government and nongovernment organizations to support research of mutual interest and of relevance to EPA's mission, explore innovative approaches for carrying out this research, and sponsor a diverse portfolio of research that alerts the agency to emerging issues and provides independent analyses of issues that the agency is currently addressing.

Does the STAR program have adequate processes to ensure that it is sponsoring high-quality and relevant research?

Finding. The procedures that STAR has established for soliciting and selecting the highest-quality research proposals compare favorably with the procedures established by other research agencies. STAR's procedures for incorporating mission relevance into its research-planning process and in the selection of proposals to fund exceed those practiced by most other agencies.

The STAR program has developed a grant-award process that compares favorably with and in some ways exceeds that in place at other agencies that have extramural research programs, such as the National Science Foundation (NSF) and the National Institute of Environmental Health Sciences. An unusually high degree of planning goes into identifying the specific research subjects to be supported. The agency also puts considerable time and thought into preparing effective research solicitations and into funding projects that are relevant to its mission and program needs.

EPA spends substantial effort in defining its research agenda, and the STAR program submits its proposed requests for applications (RFAs) to extensive review within the agency. Those efforts are intended to ensure that the research requests are focused on the issues most important to EPA.

However, the STAR program makes insufficient use of outside experts in planning its research agenda and in identifying the most important gaps in scientific knowledge. As a result, some of its early RFAs were not as well focused as they should have been.

In soliciting research proposals, STAR makes a substantial effort to reach out to the broad scientific community and to attract the most capable scientists. The RFAs are distributed widely through EPA's Web site, the *Federal Register*, announcements at professional meetings, and e-mail distributions to individuals or institutions that sign up on the STAR Web site. When the desired research is outside EPA's traditional research fields and might therefore include scientists not already involved with the agency's research program, STAR often solicits the help of other agencies that traditionally work with these scientists to ensure that they are aware of the funding opportunities.

The STAR program has established a rigorous peer-review process. Such peer-review processes are a key part of the foundation on which excellence is achieved in all research programs, including those of the National Institutes of Health (NIH) and NSF. The agency has taken effective steps to ensure that the process does not suffer from conflicts of interest and is independent. EPA provides a "firewall" that shields the peer-review process from the influence of the project officers and staff who oversee the individual-investigator, fellowship, and center awards.

Recommendation. The STAR program should continue to improve the focus of its RFAs, and when the agency does not have the capacity internally to adequately define the state of the science in a particular research field, STAR should consider greater use of external experts to assist in identifying the highest-priority research and data gaps.

Is the STAR program producing high-quality research results?

Finding. Although it is still too early for comprehensive evaluations of the research results of the STAR program, some STAR research efforts have already substantially improved the scientific foundation for decision making, and the results produced by STAR investigators have been widely published in peer-reviewed journals.

Evaluating the quality of research results is difficult and necessarily involves substantial judgment on the part of scientists with expertise in the research fields being reviewed. In addition, because of the relative youth of the STAR program, only about 40% of STAR research projects funded to date have been completed.

However, many STAR projects have resulted in articles in highly respected, peer-reviewed journals—a traditional measure of research quality. These STAR research results have already helped to improve our understanding of the causes, exposures, and effects of environmental pollution—information critical to improving the scientific foundation for decision making. For instance, STAR-funded research on particulate matter has helped to improve our understanding of the biologic mechanisms by which inhaled ambient particles cause health effects and the nature of some of those effects. These data are critical to future regulatory decisions regarding our nation's ambient air quality.

A limited bibliometric analysis by the committee indicated that the citation rate of STAR-supported research is comparable with that of research in the same fields funded by other research organizations and undertaken by other investigators. For instance, in 1997, the average number of citations of STAR-funded ecologic research was 10.5, compared with 10.3 citations of the work of all other investigators in ecology.

The committee also reviewed the backgrounds and accomplishments of a sample of STAR-funded principal investigators. The review indicated that the STAR program was funding many scientists with outstanding credentials; they have impressive research track records and are leaders in their fields; are editors of journals or officers in societies and have received awards of distinction; and were attracted to the STAR program from fields outside EPA's mission.

On the basis of the STAR program's process for awarding grants, the quality of the individuals and institutions funded by the program, and the highly competitive nature of the awards, the committee is confident that the products of STAR grants are of the highest quality.

Recommendation. EPA should continue its efforts to attract "the best and the brightest" researchers to compete for STAR funding.

Are the STAR program results useful for EPA decisions and processes?

Finding. The STAR portfolio effectively supports EPA's mission, Government Performance and Results Act goals, and ORD strategic plans. Specific STAR research projects have yielded significant new findings and knowledge critical for regulatory decision making.

The STAR program is too young to be able to document fully the extent to which its research results are being used to inform development of new regulations and environmental-management decisions. Even with respect to projects that have been completed, there is often a substantial delay between when the research results are produced and the agency decides to undertake rule-making or other actions to address the issues that were studied.

However, some STAR projects have already yielded information important for environmental decision making. For example, STAR-sponsored research in endocrine disruptors, particulate matter, and ecologic assessment has resulted in groups of peer-reviewed publications of immediate use in understanding causes, exposures, and effects of environmental pollution. Those results are directly relevant to EPA's mission to "protect human health and to safeguard the natural environment—air, water, and land—upon which life depends." For instance, STAR-funded research on particulate matter has helped to improve our understanding of the biologic mechanisms by which inhaled ambient particles cause health effects. Research on ecologic indicators has led to the development of a dynamic, economically linked model to evaluate the driving forces and ecologic consequences of land-use change.

In research fields in which EPA does not already have substantial expertise, the committee suggests that the program consider bringing in outside experts to assist in assessing the state of the science while the research program is being planned and then to synthesize the contributions of the

STAR-supported research when it has been completed. Such assessments would help EPA to target RFAs and then analyze the utility of the completed research in providing critical knowledge or otherwise strengthening and improving the foundation for environmental decision making.

To ensure the usefulness of STAR research results, it is also important for the STAR program to maintain a balanced research portfolio, including balances between "core" and "problem-driven" research and between human health and ecologic research.

Recommendation. The STAR program and ORD should develop mechanisms for documenting the extent to which its research is being used to support the agency's environmental decision making, should consider using outside experts to help document systematically the "state of the science" before research is initiated, and should synthesize the results of the research when it is completed to identify the specific contributions that STAR and ORD research has made to providing critical information.

Is the STAR program effective in providing results relevant to the appropriate audiences?

Finding. The STAR program has been commendably aggressive in experimenting with innovative approaches to communicating the results of its funded research to a wide variety of users and audiences, but its success in these efforts has been uneven.

The STAR program supports research of potential value to a variety of users and audiences, both in and outside EPA. Much of the research is aimed primarily at the scientific community and those responsible for providing technical support for environmental-management decisions. For the scientific community, the primary communication product is peer-reviewed journal articles, and the program has been successful in encouraging the preparation of these articles.

The program, however, also has other potential users, at least for the results of some of its research. They include other federal agencies; industry; state, tribal, and local governments; nonprofit environmental organizations; and international environmental agencies. The audience for some projects appears to be local communities (for instance, communities that have received Environmental Monitoring for Public Access and Community Tracking, or EMPACT, grants) or the general public; disseminating results to such audiences is much more difficult.

The STAR program has experimented aggressively with a wide variety of communication mechanisms. Information is available to the public on EPA's Web site concerning every step of the STAR process, from the initial solicitation of research proposals, through the award of grants, to the final research results. STAR researchers are required to prepare annual progress reports, which are made available to the public in summary form. The STAR program also produces several series of reports that summarize research results in selected topics. In all those efforts, the program appears to substantially exceed the dissemination efforts of most other research-sponsoring organizations, both in and outside the federal government.

Nevertheless, the STAR program could substantially improve its dissemination efforts by directing its communication efforts more effectively to specific users and audiences. The program does not always clearly identify the users and audiences for its research results. Often, the research results are produced, and then EPA assesses how to communicate them. The dissemination process would be much more effective and efficient if the potential audiences were clearly identified before the research began and if the focus were maintained throughout the research process and the preparation of reports.

In some cases, the effective dissemination of results should be primarily STAR's responsibility. In other cases, STAR's contributions will be a component of a larger research effort, and the primary dissemination responsibility should lie with ORD or EPA. In all cases, however, dissemination efforts are likely to be more effective if the intended audiences are clearly defined from the beginning of the STAR grants process.

Recommendation. The STAR program should clearly identify the intended audiences for proposed research results as early in the process as possible and should identify the audiences in RFAs. When appropriate, EPA should consider involving representatives of the intended audiences from outside the agency in helping to define the relevant research results and the strategy for their dissemination.

Should the fellowship program continue to be part of the ORD research program?

Finding. The STAR fellowship program is a valuable mechanism for enabling a continuing supply of graduate students in environmental sciences and engineering to help build a stronger scientific foundation for the nation's environmental research and management efforts.

The fellowship program was established to "encourage promising students to obtain advanced degrees and pursue careers in environmentally related fields" and to develop the next generation of environmental scientists. It is the only federal fellowship program exclusively designed for students pursuing advanced degrees in environmental sciences. It has achieved its goals, as evidenced by the extraordinary competition for the fellowships and the rigorous, independent selection process. Of the fellowship applications that STAR receives annually, only 125 fellowships, or 10% of all applicants, receive funding. Of the more than 100 former EPA fellowship recipients that were contacted by the committee, over 95% indicated high satisfaction with the program, and nearly 90% have remained in the environmental field, thus successfully contributing to the long-term program goals.

Recommendation. Given the nation's continuing need for highly qualified scientists and engineers in environmental research and management, the STAR fellowship program should be continued and funded.

Are the STAR program's funds adequate to achieve its objectives?

Finding. STAR is only able to fund less than 15% of the proposals received for its individual investigator and center grants, and its funding has not kept pace with the rate of inflation.

NIH and NSF strive to fund, on the average, 25-30% of the proposals received. STAR's budget allows it to fund only 10-15% of the proposals it receives and only about 60% of those rated "excellent" or "very good" by its independent quality peer-review panels. By that measure, STAR does not have sufficient funds to recognize all the best proposals received.

To be effective in its partnerships with other agencies, STAR must have sufficient funding to allocate to subjects of mutual interest to make it worth the extra administrative effort that partnerships require. The partnerships benefit STAR as a result of both the funds they leverage and the reputation they bring to the program.

Although the STAR program's budget grew rapidly in its first 3 years, it has not kept pace with general inflation in the last few years. That is particularly true of the STAR fellowship program. The effect of that budgetary situation is exacerbated by the fact that costs of research have outpaced general inflation for more than a decade. Therefore, at present, STAR funds buy less research than the same amount of money could have bought several years ago.

It is appropriate to consider the funding of the STAR program in the context of the overall funding for all of ORD, which also has not kept pace with inflation. STAR currently represents about 18% of ORD's total funding. The committee considers that percentage to be a reasonable recognition of the value of independent peer-reviewed research to the agency.

Recommendation. STAR program funding should be maintained at 15-20% of the overall ORD budget, even in budget-constrained times. However, budget planners should clearly recognize the constraints of not having inflation escalators to maintain the level of effort of the entire program.

How should the STAR program be evaluated?

Finding. There are no easy answers when it comes to identifying metrics for evaluating research programs, and the best approach for evaluating the STAR program is to establish a structured system of reviews by panels of experts.

The STAR program has undergone a substantial—some might say excessive—number of reviews. Most of the reviews have focused on the program's procedures; it is too early in the program's life to be able to evaluate the research products fully. Too many reviews can be disruptive to the program and can divert the program's attention and resources from its primary purpose.

The committee, in its own evaluation of STAR, assessed the quality, relevance, and performance of the program, as set forth in recent OMB research and development criteria, by using qualitative and quantitative metrics selected on the basis of its review of information available from EPA and metrics used by EPA and other organizations. That is one approach for reviewing the STAR program and similar programs. Several examples of qualitative and quantitative metrics that were used for evaluating the STAR program are these: Does the STAR program have a clearly defined plan for regular, external reviews of its research quality, and has this plan been effectively carried out? Has the program made significant contributions to advancing the state of the science in particular research topics? Does the program award grants expeditiously? Does the program have a schedule for the products it intends to produce and how well is it adhering to the schedule?

The committee's judgment is that quantitative metrics, although outwardly simpler to use, are not necessarily more informative than qualitative metrics. In some cases, quantitative metrics can be misleading, and emphasizing inappropriate metrics can distort the research outputs of a program. Qualitative metrics are less likely to have such effects, but they need to be interpreted carefully.

The committee judges that expert review by a group of people with appropriate expertise is the best method of evaluating broad research programs, such as the STAR program. Expert review is appropriate for evaluating both the processes and the products of the STAR program. The types of experts needed depend on the level of review being conducted—individual projects or programmatic levels. Both qualitative and quantitative metrics can provide valuable support for such expert reviews.

In planning for future reviews, the committee recommends that STAR and ORD consider an evaluation structure for the STAR program that has four levels: level 1 should examine the individual research projects, level 2 should focus on topics or groups of research projects on the same subject, level 3 should address the STAR program as a whole, and level 4 should tackle the question of how the STAR program relates to the broader institutions of ORD and EPA. The primary mechanism of review at levels 2-4 should be the panel of independent experts with the appropriate scientific, management, and policy backgrounds; the panels' evaluations can use metrics appropriate to the specific level of review. Such a structured review strategy could replace the number of ad hoc, unplanned, and uncoordinated reviews.

Recommendation. STAR and ORD should establish a structured program of reviews by panels of independent experts and should collect the appropriate information to support these reviews.

1

Introduction

The Environmental Protection Agency (EPA) mission is to protect human health and to safeguard the natural environment—air, water, and land—on which life depends. The *EPA Strategic Plan* (EPA 2000) emphasizes the role of science in accomplishing the agency's mission: "science is the foundation that supports all of EPA's work, providing us with the knowledge and technologies to detect, abate, and avoid environmental problems." In fact, one of the agency's goals is "sound science, improved understanding, and innovation." EPA has sought to support its goal of sound science by establishing a research program encompassing both human-health and environmental disciplines.

Several previous reports have addressed the role of science in EPA and the issue of whether a research program should be maintained within the agency (EPA 1992, 1994; NRC 1997, 2000; Powell 1999). Those reports have all stated emphatically that research is vital to the agency's mission and that EPA needs to support and maintain a strong research program. The 1992 EPA Science Advisory Board report *Safeguarding the Future: Credible Science, Credible Decisions* concluded that EPA needs its own strong science base to provide the background required for effective environmental protection programs. Similarly, *Building a Foundation for Sound Environmental Decisions*, the 1997 report of the National Research Council's Committee on Research Opportunities for EPA (NRC 1997), concluded that EPA needs a strong in-house research program. The Research Council's 2000 report *Strengthening Science at the U.S. Environmental Protection*

Agency (NRC 2000) agreed that a vigorous research program should be maintained in EPA, stating that

> moving the research program out of the agency would most likely weaken, not strengthen, the scientific foundation of EPA's decisions and actions ... An EPA devoid of a research program would not be likely to attract substantial scientific talent, and an EPA without scientific talent would be ineffective and potentially harmful to the nation (NRC 2000).

EPA'S RESEARCH PROGRAM

EPA's current research program consists of "core" and "problem-driven" research. Those terms were coined by the National Research Council committee that wrote *Building a Foundation for Sound Environmental Decisions*, which recommended that EPA's research program maintain a balance between problem-driven research, targeted at understanding and solving particular identified environmental problems and reducing the uncertainties associated with them, and core research, which aims to provide broader, more generic information to help improve understanding relevant to environmental problems for the present and the future. The report described problem-driven research as the kind of research and technical support activity that EPA's Office of Research and Development (ORD) has pursued most in the past—efforts that are driven largely by current or expected regulatory efforts of other EPA offices.

The 1997 Research Council report pointed out that the distinction at EPA between core and problem-driven research is not always clear-cut, and it is important to note that the terms are not the same as *basic* vs *applied* research, *fundamental* vs *directed* research, or *short-term* vs *long-term* research, which are typically used by other federal agencies and researchers.

Research in EPA is overseen by the Office of Research and Development (ORD), which is based at EPA headquarters in Washington, D.C. ORD's mission is to conduct leading-edge research and to foster the sound use of science and technology to fulfill EPA's mission to protect human health and safeguard the natural environment. That mission commits ORD to conduct its research in a way that will have a direct and meaningful influence on EPA's decisions and programs (EPA 2001). The *ORD Strategic Plan* (EPA 2001) defines the goals and strategies for achieving its mission-related activities.

ORD has dual roles: providing technical support for regulatory programs and acting as an independent source of scientific research and assessment (Powell 1999). ORD therefore encompasses diverse activities, including conducting an intramural R&D program, administering competitive and noncompetitive extramural R&D programs, managing some of the agency's scientific databases, providing technical support to program offices, conducting substance-specific risk assessments, and helping program offices with risk-assessment guidance (Powell 1999).

ORD comprises the National Center for Environmental Research (NCER), the National Center for Environmental Assessment, the National Exposure Research Laboratory, the National Health and Environmental Effects Research Laboratory, the National Risk Management Laboratory, the Office of Science Policy, the Office of Resources Management and Administration, and the National Homeland Security Research Center.

Users of ORD research include not only the program and regional offices of EPA but also state, tribal, local, and international government agencies that make environmental decisions; other federal agencies; academe; and the public.

EPA has a science and technology budget of $700 million, of which about $544 million goes to ORD, $106 million to the Office of Air and Radiation, and $50 million to all other programs (http://www.house.gov/science/hearings/ets02/apr23/ets_charter042302.htm). About one-third of ORD's funding is spent on intramural research, and two-thirds on extramural research (J. Puzak, EPA, Washington, D.C., personal commun., August 19, 2002). About $100 million of ORD's extramural funding is spent on the Science To Achieve Results (STAR) research grants program.

THE STAR PROGRAM

The STAR program was established in 1995 to augment EPA's research and scientific activities through coordinated funding efforts in the academic and nonprofit communities. The program was intended to ensure the best possible quality of science in fields of greatest importance to the agency. The STAR program, operated by ORD's NCER, constitutes EPA's largest single investment in extramural research. The research support awarded by the STAR program is of three main kinds: grants awarded to individual investigators or small groups of investigators, grants awarded to multidisciplinary (and sometimes multi-institutional) research centers, and fellowships to support graduate work in environmental sciences at the master's and Ph.D. levels. The program focuses specifically on meeting the research needs of EPA and is run in accordance with the *ORD Strategic Plan* (EPA 2001).

Research conducted under the STAR program covers a wide array of topics, including highly technical research on toxicology and environmental chemistry and physics, community monitoring, and socioeconomic topics. Some of the major research fields represented are air pollution, water and watersheds, ecosystem analysis, and environmental technology.

The STAR program was created in response to specific needs identified by Congress in 1994 (Senate Report 103-311). Robert Huggett, assistant administrator of ORD at the time, reorganized ORD and initiated the STAR program by reallocating $57 million in funds from other EPA research efforts.

THE NATIONAL RESEARCH COUNCIL COMMITTEE

The director of NCER approached the National Research Council's Board on Environmental Studies and Toxicology about conducting an independent assessment of the STAR research grants program. To conduct the study, the Research Council convened the Committee to Review EPA's Research Grants Program, which prepared this report. The committee's members were selected for expertise in research program administration, program evaluation, technology transfer, environmental science, risk assessment, risk management, and environmental engineering. None of the committee members was a current recipient of a STAR grant, nor did any committee member apply for a STAR grant during the course of the study.

The committee was charged with conducting a program review of the STAR competitive extramural grants program, assessing the program's scientific merit, its demonstrated or potential influence on policies and decisions, and other program benefits that are relevant to EPA's mission. It was asked specifically to examine the program's research priorities, research solicitations, peer-review process, current research projects, and results and dissemination of completed research in the context of other relevant research conducted or funded by EPA and in comparison with those of other basic and applied research grants programs.

To address its task, the committee held three public sessions in which it heard presentations from EPA officials in ORD, program offices, regional offices, and the Board of Scientific Counselors; the General Accounting Office; the National Science Foundation; the Department of Energy; the U.S. Department of Agriculture; the National Institute of Environmental Health Sciences; and academe. The committee also evaluated background information provided by EPA. Committee members held meetings and conference calls with ORD project officers, STAR grant recipients, and fellowship recipients. Committee members also attended EPA-sponsored STAR workshops and meetings. The committee was urged by ORD to develop and use metrics in its evaluation of the STAR program.

In addressing its charge, the committee was mindful of several facts. It was tasked with reviewing how well the STAR program is operating, not with revisiting the decision to establish the STAR program or assessing the overall structure of the EPA research program. Research follows a long-term course, typically requiring 3-5 years to conduct laboratory or field studies, analyze the results, and finally publish the results in peer-reviewed journals. The committee recognized that because the STAR program is relatively new, only about 40% of the research projects funded to date have been completed, and many results have not yet appeared in the published literature or been cited in regulatory documents. In the committee's evaluation, it focused on the grant program's quality, relevance, and performance in accordance with the recent Office of Management and Budget (OMB) guidelines on evaluating research programs as required by the Government Performance and Results Act; the committee considered that analyzing the STAR program in terms of OMB's criteria would provide valuable guidance to EPA (OSTP/OMB 2002). The committee used quantitative and qualitative metrics that permitted it to form judgments regarding the scientific merit of the program and its influence on the agency. Finally, the committee is aware that the STAR program has been the subject of several other recent reviews and considered them when forming its conclusions and recommendations.

ORGANIZATION OF THE REPORT

The body of this report is organized in five chapters. Chapter 2 presents an overview of the STAR program, including the components and operation of the program. Chapter 3 compares the procedural aspects of the STAR program with those of other federal competitive extramural grant programs. Chapter 4 reviews metrics and their use in evaluation of research programs. Chapter 5 presents the committee's evaluation of the STAR program, using both quantitative and qualitative metrics to assess whether the program is achieving its stated objectives. The evaluation was conducted in the context of OMB's R&D criteria: quality, relevance, and performance.

REFERENCES

EPA (U.S. Environmental Protection Agency). 1992. Safeguarding the Future: Credible Science, Credible Decisions. The Report of the Expert Panel on the Role of Science at EPA. EPA/600/9-91/050. U.S. Environmental Protection Agency, Washington, DC.

EPA (U.S. Environmental Protection Agency). 1994. Research, Development, and Technical Services at EPA: A New Beginning, Report to the Administrator. EPA/600/R-94/122. U.S. Environmental Protection Agency, Washington, DC.

EPA (U.S. Environmental Protection Agency). 2000. EPA Strategic Plan. EPA 190-R-00-002. Office of the Chief Financial Officer, U.S. Environmental Protection Agency, Washington, DC. September 2000. 104pp.

EPA (U.S. Environmental Protection Agency). 2001. ORD Strategic Plan. EPA/600/R/01/003. Office of Research and Development, U.S. Environmental Protection Agency, Washington, DC. [Online]. Available: http://www.epa.gov/ospinter/strtplan/documents/final.pdf [accessed Jan. 22, 2003].

NRC (National Research Council). 1997. Building a Foundation for Sound Environmental Decisions. Washington, DC: National Academy Press.

NRC (National Research Council). 2000. Strengthening Science at the U.S. Environmental Protection Agency, Research-Management and Peer-Review Practices. Washington, DC: National Academy Press.

OSTP/OMB (Office of Science Technology and Policy/Office of Management and Budget). 2002. FY 2004 Interagency Research and Development Priorities. Memorandum for the Heads of Executive Departments and Agencies, from John Marburger, Director, Office of Science and Technology Policy, and Mitchell Daniels, Director, Office of Management and Budget, The White House, Washington, DC. May 30, 2002.

Powell, M. 1999. Science at EPA: Information in the Regulatory Process. Washington, DC: Resources for the Future.

2

Overview of the STAR Program[1]

The Science to Achieve Results (STAR) program is the primary funding mechanism of the Environmental Protection Agency (EPA) for supporting extramural research grants and graduate fellowships in engineering and the environmental sciences. The program was established to augment EPA's research and scientific activities by funding independent but coordinated research efforts at academic and nonprofit research institutions.

Before the establishment of the STAR program, EPA supported its regulatory mission through research conducted at or sponsored by an array of laboratory and other technical facilities across the nation. The Office of Research and Development (ORD) operated 12 research laboratories, four assessment offices, and four field stations (EPA 2003a).

The ORD laboratories were headed by managers who had extensive resources within their control, including funds for supporting intramural and extramural research. There was the perception by many people that deci-

---

[1]The information in this chapter was obtained from presentations to the National Research Council committee, interviews with various EPA staff, the experience of committee members who have worked with EPA's Office of Research and Development and the STAR program, and informal communication with many persons associated with the STAR program, both in and outside the agency. Much of the information presented here is a composite of multiple observations from those diverse sources and cannot be attributed to any specific source. The committee has attempted to verify the information, but committee members have not observed most of the processes and procedures described, which generally have not been recorded in published documents.

sion making was highly decentralized; laboratory managers had substantial local autonomy and control over funding decisions. There was no coherent and transparent policy for judging and selecting proposals or cooperative agreements, and peer review, as it is commonly used in the scientific community, often was not used (Johnson 1996).

The growing U.S. environmental agenda placed an increasingly heavy burden on ORD for new research results; it was increasingly difficult for ORD to respond in a timely manner, and laboratory managers relied more heavily on contracts and cooperative agreements for meeting the demands. Before 1992, ORD funding for contracts was roughly $160 million, for cooperative agreements $100 million, and for research grants $40 million. Those funding divisions created problems related to the proper management of the research and to ensuring that the work was responsive to the needs of the program offices (Johnson 1996).

This chapter reviews the evolution of the STAR program; the components of the current program, including the research fields it covers; and the procedures for selecting research topics and awarding grants.

EVOLUTION OF THE PROGRAM

Robert Huggett, the assistant administrator of EPA for ORD, reorganized ORD and initiated the STAR program in 1995 by reallocating $57 million in funds from other ORD-sponsored research efforts (primarily the "exploratory research" program). The STAR program was assigned to one of the agency's newly established research centers, the National Center for Environmental Research and Quality Assurance, now known as the National Center for Environmental Research (NCER) (see Figure 2-1). The program's research focus has been developed specifically to meet the research needs of EPA and is run in accordance with the *ORD Strategic Plan* (EPA 2001).

Although the STAR program apparently was not established with a defined mission or set of goals, EPA has developed a set of 6 goals for the program (P. Preuss, EPA, Washington, D.C., personal commun., August 5, 2002):

- Achieve excellence in research.
- Focus on the highest-priority environmental science and engineering needs to assist EPA in its mission.

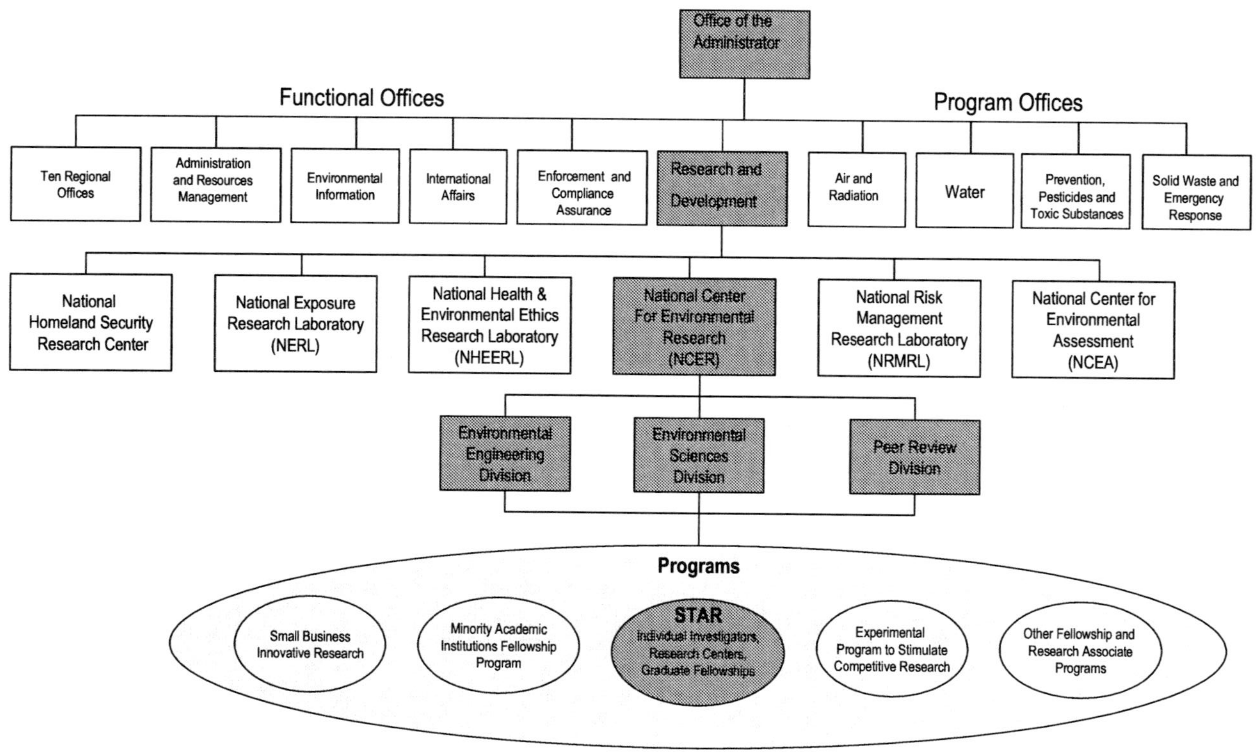

FIGURE 2-1 Location of the STAR program in the EPA hierarchy.

- Develop the next generation of environmental scientists.
- Achieve high levels of accountability and integrity.
- Form partnerships and leverage resources.
- Communicate and integrate research results.

The program began with three components: focused requests for grant applications, an exploratory research grants program (which invited grant applications to conduct exploratory research in environmental physics, chemistry, and biology without designating particular program foci), and a graduate fellowship program (EPA 1996a). During the intervening years, the components and management of the program have changed in response to changing agency needs, experience gained in operating the program, and external reviews.

The program grew rapidly during its first 4 years, but its funding has since remained relatively constant, with the total STAR budget fluctuating around $100 million per year, as indicated in Figure 2-2. In its initial year, the STAR program accounted for 11% of total R&D expenditures. Over the intervening years, total expenditures for ORD have fluctuated between $500 and $600 million, and the STAR program now accounts for about 18% percent of the ORD total.

Accounting for changes in prices, of course, the value of the STAR funds has increased less than the expenditures. For instance, when deflated by the consumer price index, the value of the STAR grants was about the same in 2002 as in 1997, 2 years after the program began. The problem of increasing costs is most noticeable in the fellowship component of the program. Tuition costs have typically been increasing at twice the rate of inflation over the last decade and in some regions have almost doubled (College Board 2001). EPA pays for the increases as long as tuition does not exceed $12,000 (including fees), but it has not adjusted the cap to reflect increasing tuition costs since the program began.[2] At high-cost institutions, the fellowships may therefore be much less adequate than they were when the program began (although still larger than those offered by many other fellowship programs).

During its 8-year life, the STAR program has evolved in several ways. Some of the changes have accompanied changes in the program's funding. The average size of the individual investigator and center grants has in-

---

[2]As this report was being prepared, EPA reported that it is reviewing this issue (J. Puzak, EPA, personal commun., October 4, 2002).

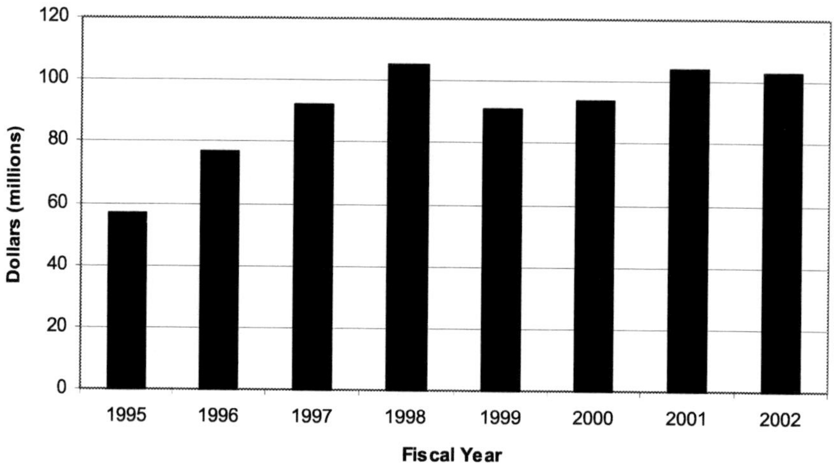

FIGURE 2-2 STAR program appropriations. Source: P. Preuss, EPA, presentation to National Research Council committee, March 18, 2002.

creased from $289,000 in FY 1995 to $743,000 in FY 2001 (J. Puzak, EPA, Washington, D.C., unpublished material, 2002). As EPA allocated more funds to the STAR program, it was able to induce other agencies with similar interests to enter into partnerships and provide supplementary funds (Figure 2-3). The agency has increased the funding of research fellowships approximately in proportion to the total funding for the STAR program (see Figure 2-4), and this has resulted in an increase in the number of fellowships awarded. However, because the size of EPA's grants has increased, the increased funding has not resulted in an increase in the number of grants awarded.

The program has evolved in other respects. Part of the evolution has been in response to changing agency research priorities. The amount of money allocated to the exploratory grants program has diminished; the emphasis has shifted to "focused" research solicitations, although some of these solicitations may also support some very basic, or "core," research efforts.

EPA has modified, improved, and strengthened important elements of the program as it has gained experience in managing a peer-reviewed, competitive research program. For instance, the improved quality of the peer-

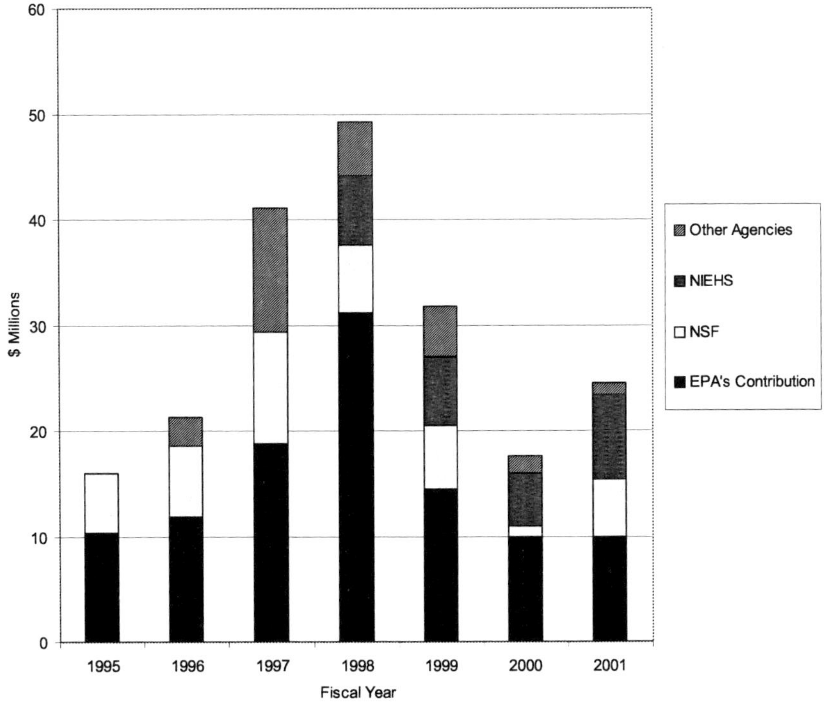

FIGURE 2-3 Funding for STAR partnerships. NIEHS, National Institute of Environmental Health Sciences; NSF, National Science Foundation. Source: P. Preuss, EPA, presentation to National Research Council committee, March 18, 2002.

review panels reflects the agency's experience in selecting panel members, identifying possible conflicts of interest, and managing an independent peer-review process. The agency has also substantially improved the information provided to the public on the substance and progress of individual grants.

Similarly, many of the early requests for applications (RFAs) were quite general, and as a result the proposals submitted were not always well focused on the agency's specific research needs. Where the agency considered this to be a problem, it has made the RFAs more focused. That also was apparently one of the reasons for reducing the exploratory research grants program.The agency recognized that some research topics could be

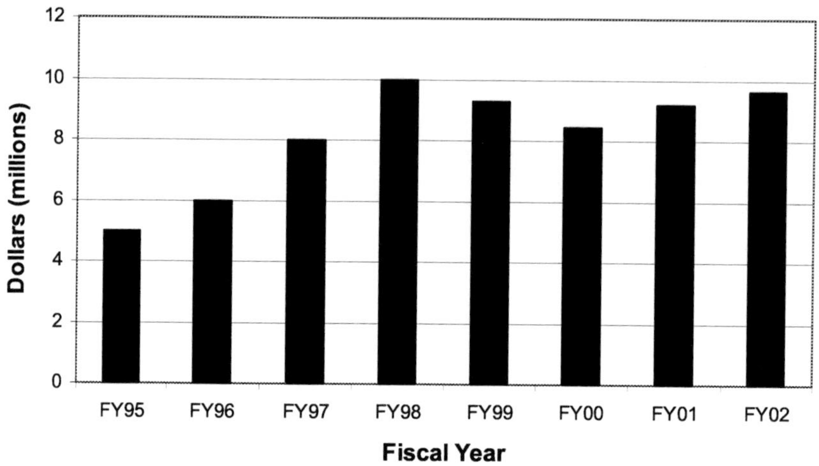

FIGURE 2-4 STAR graduate fellowship program obligations. Source: J. Puzak, EPA, Washington, D.C., unpublished material, 2002.

addressed better by interdisciplinary teams of researchers than by individual researchers working under uncoordinated grants. For that reason, the STAR program began to provide more funding to interdisciplinary "research centers."[3]

Modifications have occurred as the program has matured. In the first years, the focus understandably was on establishing and improving the grant-making process. The agency has since increased emphasis on reviewing the progress of individual research efforts, encouraging coordination among researchers, and stimulating cooperation between intramural and extramural research efforts. The primary mechanism for accomplishing those ends is the "progress review" meetings, which include all the principal investigators working on a particular topic.

Similarly, now that some of the early research projects have been completed, the agency is increasing its emphasis on developing effective ways of communicating research results to potential users. Some of the specific communication efforts are described later in this chapter.

---

[3]EPA had funded some research centers before the STAR program was begun, and those were incorporated into the STAR program in FY 1997 (EPA 1996a).

Finally, the agency has modified the STAR program in response to its many external reviews. Those reviews and the agency's responses to them are summarized in Appendix B of this report.

COMPONENTS OF THE PROGRAM

The STAR program has three main components: individual investigator awards, research centers, and student fellowships.

Individual Investigator Awards

Individual investigator awards provide funding to individual investigators or small teams of cooperating investigators who propose to conduct research on topics identified by the agency. The proposals are investigator-initiated through universities, colleges, and nonprofit research institutions. Awards are generally for 3 years and for about $50,000 to $1,000,000. In most years, the STAR program has funded 170-200 individual investigator awards (EPA 2003b).

In some cases, the RFAs are issued jointly by EPA and one or more other federal agencies or organizations—most significantly, the National Science Foundation (NSF) and the Department of Health and Human Services—interested in similar research issues.[4] In any year, an additional 30-50 grants are awarded by the other agencies and organizations for the joint RFAs; these awards supplement the STAR awards and are not included in the STAR program statistics.

Research Centers

The research centers fund multidisciplinary efforts involving a number of scientists working in complementary fields. The multidisciplinary aspect

---

[4]Cooperating agencies have included the Department of Energy, the Office of Naval Research, the National Aeronautics and Space Administration, the National Oceanic and Atmospheric Administration, the U.S. Department of Agriculture, the National Institute for Occupational Safety and Health, the National Institute of Environmental Health Sciences, the Department of the Interior, the American Waterworks Research Foundation, and the Association of California Water Agencies.

of the centers allows research programs to incorporate, for instance, exposure assessment and health-effects research with validation of risk- management and health-prevention strategies. Several research organizations or institutions may be involved in one center. Most centers are funded for 5 years, and the amount of the award typically exceeds $5 million over the life of the grant. Centers can also be jointly funded. For instance, the National Institute of Environmental Health Sciences (NIEHS) and EPA jointly fund the Centers of Excellence in Children's Environmental Health and Disease Prevention Research. Figure 2-5 shows the amount of funding obligated for research centers and the shift from individual investigator awards to research centers. In FY 1995, all the research funds were for individual investigator awards; in FY 2001, about one-third of the STAR grant obligations went to support research centers.

Fellowships

The STAR Graduate Fellowship Program was established to "encourage promising students to obtain advanced degrees and pursue careers in environmentally related fields" (EPA 2002a). About 125 fellowships are awarded each year, although the Office of Management and Budget (OMB) proposed eliminating funding for this program in FY 2002 (P. Preuss, EPA, personal commun., March 18, 2002). The program is the only federal fellowship program designed exclusively for students pursuing advanced degrees in environmental sciences.[5]

The STAR fellowships provide more financial support than most other fellowships (Hogue 2002). For instance, NSF fellowships offer a total of $27,300 annually to doctoral students for 3 years. STAR fellowships provide a total of up to $34,000 per year and are available to both master's and doctoral students. Of the total amount, $17,000 can be used for a stipend,

---

[5]Fellowships are available to graduate students in environmental engineering, atmospheric sciences, chemistry and materials science, geology (including geochemistry and geophysics), economics (including market incentives and health and ecosystem valuation), geography, genetics (including genomics, proteomics, and bioinformatics), microbiology, public-health sciences (including epidemiology, exposure assessment, biostatistics, and health risk assessment), toxicology, aquatic ecology and ecosystems, oceanography and coastal processes, entomology, forestry, zoology, terrestrial ecology and ecosystems, and ecologic risk assessment (EPA 2002f).

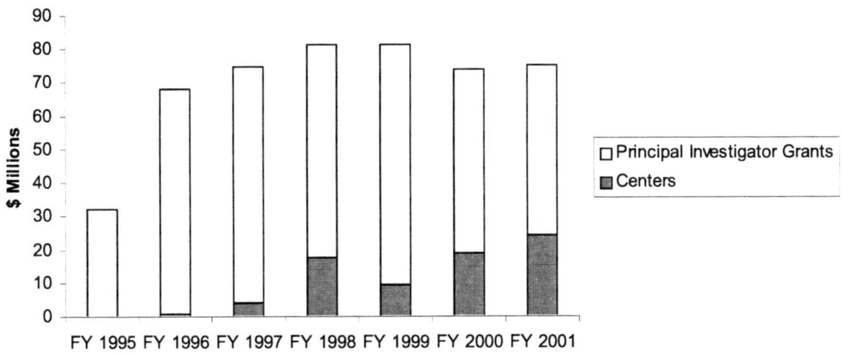

FIGURE 2-5 Grant obligations to individual investigators and research centers. Source: J. Puzak, EPA, Washington, D.C., unpublished material, 2002.

\$12,000 for tuition and fees, and \$5,000 for other authorized expenses. Students enrolled in master's programs may receive up to 2 years of support and doctoral students may receive up to 3 years of support (EPA 2003c).

In addition to contributing to the STAR goal of "develop[ing] the next generation of environmental scientists," EPA hopes that these fellowships will assist the agency in its goal of "recruiting and retaining the next generation of well trained and highly qualified scientists" (EPA 2001; Preuss 2002a). That is an increasingly important issue for the agency in that it foresees the retirement of more than the average number of its scientific staff in the near future.

STAR fellowships are highly competitive: only 10% of applicants receive funding (NCSE 2003). Prospective applicants are evaluated on the basis of rigorous peer review, academic and employment records, and potential. Fellowships have remained constant at about 10% of STAR funding. However, the program solicited no fellowship applications in FY 2003, because OMB eliminated STAR's funding for this program.

RESEARCH FIELDS

The STAR program has been involved in a wide variety of research since its inception. Table 2-1 indicates the numbers of awards and amounts of support the STAR program has "committed" to different topics during its first 7 years of operation.

TABLE 2-1 Annual Commitments by Topic[a] (in Millions of Dollars)

	FY 1995		FY 1996		FY 1997		FY 1998		FY 1999		FY 2000		FY 2001	
	No.	$	No.	$	No.	$	No.	$	No.	$	No.	$	No.	$
AIR														
Chemistry and physics	11	3.1	11	3.7			10	4.0						
Ozone	1	0.4	12	5.1	7	10.5								
Air toxics			6	3.0	3	1.4	5	2.3	4	1.7				
Particulates	2	1.0	6	1.8	9	3.7	14	44.9	8	4.5			4	3.8
Miscellaneous	9	4.1	3	1.5					5	1.7				
Subtotal	**23**	**8.6**	**38**	**15.1**	**19**	**15.6**	**29**	**51.2**	**17**	**7.9**	**0**	**0.0**	**4**	**3.8**
WATER AND WATERSHEDS														
Water and watersheds	23	9.0	11	8.8	10	8.1	9	6.5	8	6.6				
Drinking water			6	2.0	8	3.6	10	3.5	12	5.2	12	6.0	3	3.2
Contaminated sediments			3	1.7	5	1.8								
Algae blooms					4	1.0	7	3.0	2	0.9			9	3.1
Miscellaneous	14	4.4	6	1.9					9	7.0	2	0.7		
Subtotal	**37**	**13.4**	**26**	**14.4**	**27**	**14.5**	**26**	**13.0**	**31**	**19.7**	**14**	**6.7**	**12**	**6.3**

HUMAN HEALTH														
Children's health			5	3.0			8	27.1	9	5.3	5	4.1	10	18.3
Endocrine disruptors			9	4.0	15	6.3			9	6.4			4	7.9
Miscellaneous	8	3.1	9	4.0	21	12.3	9	3.6	2	6.3			2	1.9
Subtotal	**8**	**3.1**	**23**	**11.0**	**36**	**18.6**	**17**	**30.7**	**20**	**18.0**	**5**	**4.1**	**16**	**28.1**
ECOSYSTEMS														
Environmental biology	27	8.6	13	4.5	14	4.4	17	5.0						
Regional assessments			10	8.2			4	1.6	6	5.7	3	1.2		
Ecological indicators					9	6.7	22	12.8	8	5.1	8	28.1	3	7.6
Miscellaneous					7	4.4							9	3.4
Subtotal	**27**	**8.6**	**23**	**12.7**	**30**	**15.5**	**43**	**19.4**	**14**	**10.8**	**11**	**29.3**	**12**	**11.0**
TECHNOLOGY														
Sustainable development	8	1.5	13	3.2	11	3.1	14	3.5	40	10.8			15	4.8
Exploratory engineering	19	4.4	5	1.8	11	3.2	9	2.7	10	2.2				

(Continued)

TABLE 2-1 *Continued*

	FY 1995		FY 1996		FY 1997		FY 1998		FY 1999		FY 2000		FY 2001	
	No.	$	No.	$	No.	$	No.	$	No.	$	No.	$	No.	$
Bioremediation			5	2.4	5	2.1	3	1.3					4	1.5
Advanced computing			16	8.6					7	5.1				
Miscellaneous			14	4.5										
Subtotal	**27**	**5.9**	**53**	**20.5**	**27**	**8.4**	**26**	**7.5**	**57**	**18.1**	**0**	**0**	**19**	**6.3**
SOCIOECONOMIC														
Valuation and decision-making	14	2.0	7	1.5	8	1.5	11	1.9	9	1.7	1	0.3	7	2.1
Incentives	18	3.5									13	2.8	11	2.3
Community monitoring							8	3.6	8	3.5	9	3.1	8	2.8
Miscellaneous	9	1.9			3	0.7								
Subtotal	**41**	**7.4**	**7**	**1.5**	**11**	**2.2**	**19**	**5.5**	**17**	**5.2**	**23**	**6.2**	**26**	**7.2**
OTHER														
Futures	6	1.3							8	1.0	3	1.0	29	11.6
Global and climate change	7	2.7	5	8.2	1	0.3			6	5.3	4	5.6		
Environmental statistics			2	5.8			4	1.1	3	0.7	2	0.8	1	6.3

Environmental chemistry and physics					21	5.3	16	5.5	13	2.6				
Chemical mixtures							4	3.1	2	1.2	6	3.6		
Hazardous waste													5	11.3
Miscellaneous			18	7.8					2	1.6	8	3.2	2	1.3
Subtotal	**13**	**4**	**25**	**21.8**	**22**	**5.6**	**24**	**9.7**	**34**	**12.4**	**23**	**14.2**	**37**	**30.5**
TOTAL	**176**	**51**	**195**	**97.0**	**172**	**80.4**	**184**	**137.0**	**190**	**92.1**	**76**	**60.5**	**126**	**93.2**

[a]Commitments are obligations during a fiscal year plus promises for future obligations if Congress appropriates sufficient funds. The topics listed are not necessarily those identified by the STAR program. The committee developed thsi list after reviewing the full list of awards that the program made to indicate generally how the STAR program has been allocating its funds.
Source: J. Puzak, EPA, Washington, D.C., unpublished material, 2002.

In some instances, particularly in the case of research centers, the commitments are for support in future years, and the funds are not allocated in the year when the commitments are made. Thus, the amount of support "committed" can differ substantially from the amount appropriated to or obligated by the program, the more common measures of government budgetary activities.[6] In FY 1998, for instance, the program made commitments totaling $137 million, including sizable commitments to a number of new research centers. That amount substantially exceeded the amount of money appropriated for the program or obligated or expended by the program during that year. Because the commitments had to be honored by obligations in later years, the amounts committed in those years (for example, FY 2000) fell below the amounts appropriated and obligated.

In FY 2000 and 2001, the largest commitments were made in drinking water, children's health, endocrine disruptors, ecologic indicators, environmental futures, global and climate change, environmental statistics, and hazardous waste. However, substantial work was also being supported in other topics, such as the health effects of fine particles, which had received sizable commitments with multiyear obligations in earlier years.

Individual investigator grants have been awarded in every topic. Five kinds of research centers have been funded: five Airborne Particulate Matter Centers, 12 Centers of Excellence in Children's Environmental Health and Disease Prevention Research, five Hazardous Substances Research Centers, five Estuarine and Great Lakes Program Centers, and four Statistics Centers.

In much of its analysis, the committee focused on three particular research topics: particulate matter as an air pollutant, ecologic indicators, and endocrine disruptors. Looking at those in more detail may be helpful in understanding how the STAR program operates; however, these three examples may not necessarily be representative of the other research conducted within the STAR program. The committee selected them because the STAR program had provided them with substantial support in recent years, they represent a mix of small and large grants and of individual investigator awards and research centers, some committee members had a substantial familiarity with the fields and the work being done by STAR

---

[6]Appropriations and obligations have been used in most of the other diagrams and budgetary discussions in this chapter. Because the government's budget controls focus on obligations, that is usually what is meant by government "funding."

grantees,[7] and substantial information on the programs was available to the committee.

Particulate Matter

The Clean Air Act charges EPA with conducting research and developing and implementing regulations to control criteria air pollutants that have the potential to affect human health and welfare adversely, and it specifically identifies particulate matter (PM) as one of the pollutants. EPA has been conducting and sponsoring research on particles since the agency's beginning. Following the publication of the 1997 criteria document for particulate matter, a fine particle standard was established, and many questions were raised that required additional research for the subsequent criteria document to be issued 5 years later. The STAR research program on PM has enabled EPA to respond to this need more comprehensively than it could by just using its intramural program. STAR research has focused on the human health effects of PM less than 2.5 μm in diameter. Table 2-2 shows the number of RFAs, number of investigator-initiated awards, and funding levels for PM research supported by the STAR program from FY 1995 through FY 2001.

The STAR PM research program includes both individual-investigator awards and research centers. Although some PM research was funded as early as 1995, STAR issued its first RFA specifically for investigator-initiated research on PM in 1997. The original focus of the program included research on the causal mechanisms of PM toxicity; intermediate biologic end points thought to be related to morbidity; ultrafine, fine, and coarse particles; exposure assessment for PM and associated copollutants; and the composition of components of PM, such as organic compounds (nonvolatile and semivolatile) and biologics.

In addition to examining health effects of and exposures to PM, several RFAs have examined the chemistry and physics of PM, modeling, ambient measurement and analysis, and emissions. The RFAs have become more specific and focused as EPA has learned that more narrowly defined RFAs are more able to advance the state of the science (S. Katz, EPA, personal commun., August 8, 2002).

---

[7]However, no committee member had received STAR grants for research in these or any other topics.

TABLE 2-2 Commitments Made by the STAR Program for Research on Particulate Matter

	Individual Investigator Awards					Research Centers			
	No.	No.	Size of Awards, $1,000s			No.	Size of Awards, $1,000s		
Fiscal Year	RFAs	Grants	Average	Low	High	Grants[a]	Average	Low	High
1995[b]	2	14	378	172	590	0	—	—	—
1996[b]	3	8	415	333	521	0	—	—	—
1997	2	9	423	211	601	0	—	—	—
1998	1	9	530	199	736	5	8,226	7,747	8,716
1999	1	8	565	319	764	0	—	—	—
2000	0	0	—	—	—	0	—	—	—
2001	1	4	950	833	1,239	0	—	—	—

[a]Only one RFA was issued for research centers.
[b]In FY 1995 and 1996, no RFAs were issued specifically for research on particles, but some particle research grants were awarded under more general RFAs.
Source: J. Puzak, EPA, Washington, D.C., unpublished material, 2002.

EPA issued an RFA to launch five PM centers in 1998 as recommended by the National Research Council report *Research Priorities for Airborne Particulate Matter: I. Immediate Priorities and Long-Range Research Portfolio* (NRC 1998) and in accordance with a congressional mandate. The PM research centers program provides about $8 million per year for 5 years—roughly $1.5-1.8 million per year per center. The PM centers were established to "advance the understanding of PM health effects, how they occur, and improve understanding of populations who are susceptible to health effects from exposure" (EPA 2002b). The centers provide an opportunity for interdisciplinary research that allows for the leveraging of resources and the forming of partnerships in advancing the understanding of the health effects of PM. EPA established the following four priorities for the PM centers (EPA 2002b):

- "Exposure: Improve assessments of personal exposures to PM in normal human populations and in sensitive populations (that is, the elderly, individuals with respiratory or cardiovascular disease, and children).
- Dosimetry and Modeling: Develop new models regarding the amount of particulate matter deposited in the lungs of exposed individuals. This is critical in understanding the relationships between individual exposure and health responses of sensitive populations.
- Toxicology: Identify which constituents or properties of PM are most responsible for human health effects and how these effects occur. Reducing uncertainty in this area is important for human health risk assessment.
- Epidemiology: Improve understanding of which groups are particularly susceptible to health effects from PM exposure" (EPA 2002b).

The following are the five PM centers funded by STAR:

- Harvard University PM Center, focusing on ambient particle health effects, exposure, susceptibility, and mechanisms.
- New York University PM Center, focusing on the health risks posed by PM components.
- Northwest PM Center (includes the University of Washington and Washington State University), focusing on combustion-derived fine-particle composition, exposure, and health effects.
- Rochester PM Center, focusing on characterization, health effects, and pathophysiologic mechanisms of ultrafine particles.
- Southern California PM Center (includes University of California, Los Angeles; the California Institute of Technology; the Rancho Los Ami-

gos Medical Center; the University of California, Irvine; the University of California, Riverside; and the University of Southern California), focusing on sources of exposure to, and health effects of PM.

In addition to the PM research funded through the STAR program, EPA conducts PM research in other parts of ORD—including the National Exposure Research Laboratory (NERL), the National Health and Environmental Effects Research Laboratory (NHEERL), and the National Risk Management Research Laboratory (NRMRL)—and in the Office of Air and Radiation. PM research planning and coordination in EPA is guided by the recommendations of the National Research Council Committee on Research Priorities for Airborne Particulate Matter and EPA's multiyear plan for PM. A PM program manager assists in coordinating research across the agency through a research coordinating team.

Aside from EPA's PM research program, a workgroup under the Committee on Environment and Natural Resources coordinates PM research with other federal agencies, including NIEHS, the Department of Energy, the National Oceanic and Atmospheric Administration (NOAA), the National Institute of Standards and Technology, and the National Institutes of Health. EPA has not funded PM research jointly with other agencies. However, EPA issued an RFA on the role of PM air pollution in cardiovascular illness and mortality in 2002 and solicited the assistance of NIEHS and the National Heart, Lung, and Blood Institute (NHLBI) in advertising it to a larger audience of cardiovascular researchers who might otherwise not be aware of EPA research solicitations. In August 2002, EPA held a workshop that was cosponsored by NIEHS, NHLBI, and others to gather input from the environmental health science and cardiovascular research communities on the most appropriate and productive directions for research in environmentally related cardiovascular disease. EPA is expecting more joint collaborations in the future.

Ecologic Indicators

The ecologic-indicators program is part of EPA's Environmental Monitoring and Assessment Program (EMAP), which was established to develop tools to monitor and assess the ecologic health of the nation's environmental resources. The indicators are ultimately intended for the development of biologic criteria for use in national assessments of ecologic conditions and for assessments of ecologic conditions on regional and watershed scales. When the STAR program was established, $12 million was transferred from the EMAP budget to support the development of environmental indicators, and EMAP funds have continued to support this research program.

ORD's *Ecological Research Strategy* defined ecologic indicators as "any expression of the environment that quantitatively estimates the condition of the ecological resource, the magnitude of the stress, the exposure of the biological components to stress, or the amount of change in the condition" (EPA 1998). The STAR program has been involved in ecologic indicators and assessment research since FY 1997. It consists primarily of core research, focusing on the development of indicators that integrate resource types, incorporate multiple levels of biologic organization, and address multiple spatial scales.

The ecologic-indicators program has three research objectives:

- "to stimulate the development, evaluation, and integration of indicators, suites of indicators, indices, and models to improve local, regional, national, and global monitoring and assessment of ecological integrity and sustainability;
- to develop indicators of functional processes that contribute to ecological integrity and sustainability; and
- to develop indicators that identify effects of particular stressors of ecological integrity and sustainability. This program includes initiated research awards and research centers" (EPA 1997).

Table 2-3 shows the number of RFAs, number of awards, and funding levels for ecologic-indicator research supported by the STAR program from FY 1995 through FY 2001.

The 1997 RFA solicited proposals on the development of techniques and indicators that characterize and quantify the integrity and sustainability of ecosystems at local, regional, national, and global scales. One of the 1998 RFAs focused on the development of research and monitoring programs using ecologic indicators and investigating the ecologic effects of environmental stressors at pilot sites around U.S. marine and coastal sites. The RFA was issued jointly with NOAA and the National Aeronautics and Space Administration (NASA). NOAA was interested in monitoring coastal ecosystems, and NASA in developing remote-sensing capabilities. The other 1998 RFA identified three priorities:

- "The development of landscape characterization indicators that incorporate multiple resources and spatial scales. Indicators that are useful

TABLE 2-3 Commitments Made by the STAR Program for Research on Ecologic Indicators

	Individual Investigator Awards					Research Centers			
	No.	No.	Size of Awards, $1,000s			No.	Size of Awards, $1,000s		
Fiscal Year	RFAs[a]	Grants[a]	Average	Low	High	Grants[a]	Average	Low	High
1995	0	0	—	—	—	0	—	—	—
1996	0	0	—	—	—	0	—	—	—
1997	1	9	742	295	1,300	0	—	—	—
1998	2	22	584	197	898	0	—	—	—
1999	1	8	634	224	895	0	—	—	—
2000	0	0	—	—	—	4	5,953	5,812	6,000
2001	0	0	—	—	—	1	5,901	5,901	5,901

[a]Only one RFA was issued for research centers. Four centers were awarded grants from this RFA in FY 2000 and one was awarded a grant in FY 2001.

Source: J. Puzak, EPA, Washington, D.C., unpublished material, 2002.

at regional and broader scales are emphasized over those intended primarily for local use.

- The development of indicators that span multiple resource types (e.g., forests, streams, wetlands, estuaries, rangelands). Any indicator that incorporates or integrates multiple scales and multiple levels of biological organization within the context of spanning multiple resources is also emphasized.
- The development of indicators within a single resource type (for example, forests, streams, wetlands, estuaries, rangelands) that link different levels of biological organization or multiple spatial scales. The opportunity to apply cellular and molecular genetic techniques to address genetic diversity in conjunction with other levels of biological organization and multiple spatial scales is emphasized" (EPA 1998).

The 1999 RFA emphasized indicators that "(1) integrate between or among resource types, (2) incorporate multiple levels of biological organization (gene, organism, population, community, landscape), and (3) address multiple spatial scales (local, watershed, regional, national, global)" (EPA 1999). The 2000 and 2001 RFAs solicited proposals related to the development of classification schemes and associated reference conditions for use in the application of biocriteria to specific aquatic resources, such as the Great Lakes, or to ecosystems composed of wetlands, large rivers, ephemeral systems, reservoirs, lakes, streams, estuaries, near-shore coastal environments, and coral reef communities.[8]

Most of EPA research on the development of ecologic indicators is conducted through the STAR program. The laboratories in ORD are involved primarily in the "proof of concept" or "implementation" of the indicators in the field (B. Levinson, EPA, personal commun., August 5, 2002). Research on the development of ecologic indicators is not typically conducted elsewhere in EPA.

Endocrine Disruptors

ORD first listed endocrine disruptors as a high-priority research topic

---

[8]EPA considers the centers to be part of the Estuarine and Great Lakes program, but the committee has included them with the ecologic indicators program because they are intended to make substantial use of ecological indicators for assessing the environmental health of ecosystems on which the program is focusing.

in its research plan in December 1995 and later developed an endocrine disruptors research plan that focused on: biologic-effects studies, exposure studies, and studies on the linkage between exposure and effects. The goal of the endocrine disruptors research program is "to evaluate potential health effects associated with endocrine disruptors and to determine the extent of current exposures" (EPA 1996b). The endocrine-disruptors research funded through the STAR program addresses the effects of endocrine disruptors on both human health and the environment. EPA is one of the few agencies that provide extramural funding for examining the effects of endocrine disruptors on wildlife (E. Francis, EPA, Washington, D.C., personal commun., August 6, 2002).

The STAR endocrine-disruptors research program consists only of individual investigator-initiated awards. EPA issued the first RFA in 1996 and later ones in 1997, 1999, and 2001 (see Table 2-4). The RFA in 1996 was very broad, requesting grant proposals in human health effects, ecologic effects, human exposure evaluations, and ecologic exposure evaluations. The 1997 RFA also covered a broad array of topics, but the RFAs have since become more focused.

EPA's endocrine-disruptor research is coordinated with a substantial amount of associated research sponsored or conducted by other agencies; some of this research is sponsored jointly with EPA. For instance, in 1999, EPA funded a solicitation jointly with the Department of the Interior and NOAA; the RFA involved both the population-level effects of endocrine disruptors in wildlife and the human health effects of endocrine disruptors during development. The 2001 RFA was funded jointly by NIEHS, the National Cancer Institute, and the National Institute for Occupational Safety and Health. It requested applications pertaining to epidemiologic studies on the effects of exposure to endocrine disruptors, particularly reproductive and developmental effects in humans.

In addition to the STAR program, several ORD laboratories are conducting research on endocrine disruptors. NHEERL conducts a wide array of research on the topic, including developing protocols for screening and testing, determining dose-response curves, and investigating modes of action. NRMRL, NERL, and the National Center for Environmental Assessment are also involved in endocrine-disruptor research. However, those research efforts are different from the research sponsored by the STAR program, which is used to identify data gaps and fund complementary research. For instance, the STAR program is attempting to fill information gaps concerning exposure to endocrine disruptors and epidemiologic research because the ORD laboratories have little capacity to undertake such research themselves.

TABLE 2-4 Commitments Made by the STAR Program for Research on Endocrine Disruptors

	Individual Investigator Awards					Research Centers			
			Size of Awards, $1,000s				Size of Awards, $1,000s		
Fiscal Year	No. RFAs	No. Grants	Average	Low	High	No. Grants	Average	Low	High
1995	0	—	—	—	—	0	—	—	—
1996	1	9	430	190	599	0	—	—	—
1997	1	15	423	159	598	0	—	—	—
1998	0	—	—	—	—	0	—	—	—
1999	2	9	709	257	1,265	0	—	—	—
2000	0	—	—	—	—	0	—	—	—
2001	1	4	1,982	967	2,779	0	—	—	—

Source: J. Puzak, EPA, Washington, D.C., unpublished material, 2002.

OPERATION OF THE PROGRAM

The STAR program may receive over 1,000 grant proposals per year (not counting fellowship applications), 10-15% of which are likely to be funded (EPA/SAB/BOSC 2000). Because almost all the grants are for a duration of more than 1 year, the STAR staff is responsible for managing 650-750 research grants at any time (EPA 2002a). The grant award process, which is the same for both individual investigator awards and research centers, is depicted in Figure 2-6.

Planning

The STAR grants process begins in the ORD research planning process. The STAR program was designed, from its beginning, to be integrated into and complement EPA's overall research and development program. The integration begins with the preparation of multiyear research plans for particular subjects of interest to the agency—for instance, the health effects of fine PM or ecologic indicators of water quality. For the most part, the plans are internal working documents and are not available to the public. Multiyear plans become public if they are submitted to EPA's Science Advisory Board for review. That was done with the multiyear plans for water quality and pollution prevention in 2001 (EPA/SAB 2001a).

The plans, which typically have a horizon of 5-7 years, are developed by an intra-agency team that includes representatives of the various EPA research laboratories and centers, interested program offices, and regional offices. Each plan is developed under the leadership of a program manager who is employed in some unit of ORD. Because they are internal EPA working documents, there typically is no input or review by other agencies, EPA advisory committees, or other members of the public. The plans set forth the research agenda, including topics and schedules, and assign responsibilities for undertaking the research to different ORD units. One of those units is NCER, which manages the STAR program. The research topics assigned to STAR are those in which ORD lacks expertise or is otherwise not capable of carrying out the required research on the schedule set forth in the plan. The plans are reviewed, and if appropriate modified, by the responsible intra-agency team annually or, if necessary, more frequently. New information, scheduling changes, or shifts in priorities are some of the reasons for modifying plans.

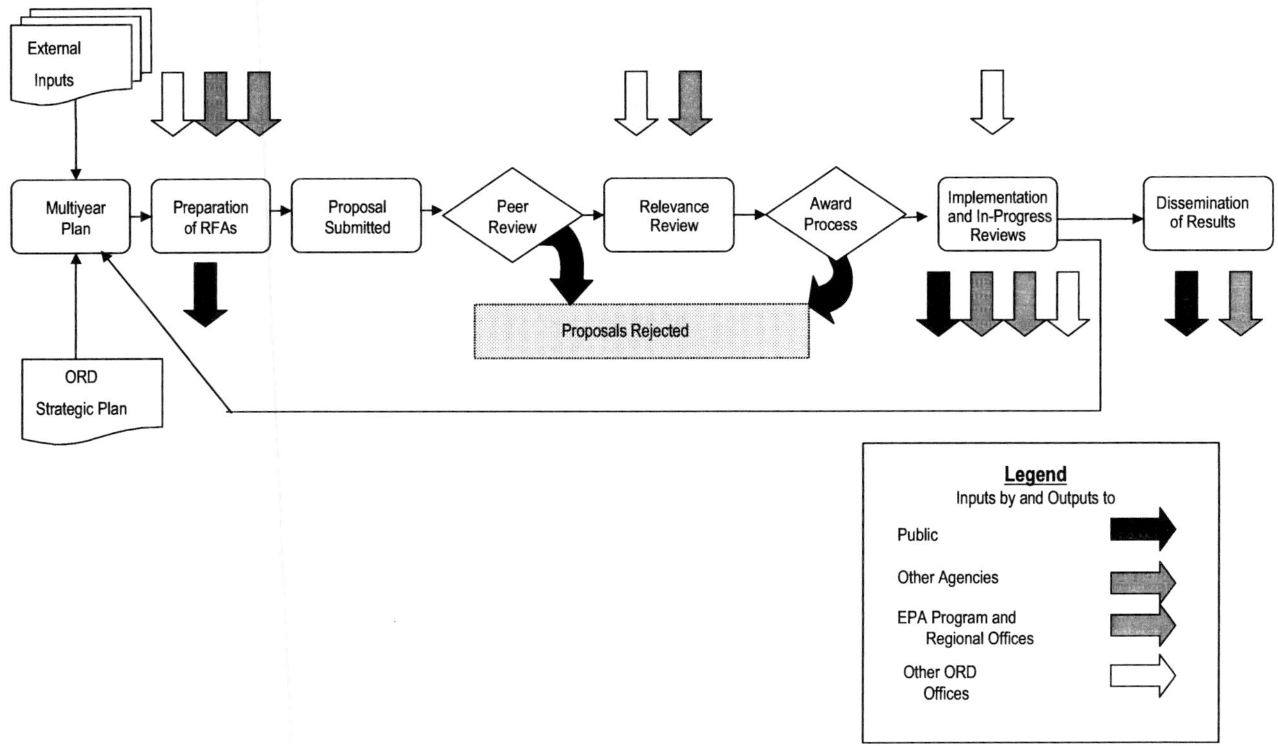

FIGURE 2-6 Operation of the STAR grants program.

Opportunity Communication

The primary vehicles for communicating STAR research opportunities are the RFAs prepared by STAR staff in cooperation with representatives of other EPA offices. They describe (typically in 2-10 pages) the specific subjects in which the agency intends to provide STAR grants. The topics selected for RFAs are those previously identified in the agency's multiyear research plans. If the research is to be sponsored jointly with other agencies or organizations, they will also be involved in preparing the RFA. The RFAs specify the type of research desired, whether grants will be awarded to individuals or research centers, the expected size of grant awards, and the deadline for proposals.

The STAR project officer and the leader of the appropriate research coordinating team use the multiyear plans to guide their preparation of RFAs. The draft RFAs are reviewed and approved by the appropriate program offices and representatives of regional offices and other ORD units to ensure that they are consistent with the intention of the multiyear plan. If the research is to be funded in part by other agencies, they are also involved in the preparation, review, and approval of the RFA.

RFAs are announced in the *Federal Register*, posted on the NCER Web site for at least 90 days, and e-mailed to individuals and institutions that have indicated an interest in receiving them by signing up on the STAR Web site. In addition, STAR and other EPA staff, or a cooperating agency or organization, may advertise RFAs at professional meetings and conferences and send copies to institutions or individuals they know to be conducting research in the subjects of the RFAs. To eliminate the appearance of potential conflicts of interest, the STAR program does not accept "preproposals."

Proposal Review

NCER has established an independent Peer Review Division for receiving, processing, and conducting initial reviews of proposals submitted in response to RFAs. The division's major responsibility is to organize and manage the peer-review panels responsible for evaluating the scientific merit of the proposals. The panels are composed of scientists who have undergone rigorous conflict-of-interest review. In some cases, because so many U.S. researchers are involved with EPA or the other institutions submitting the proposals, the conflict-of-interest requirements require the agency to identify Canadian, European, or other foreign scientists to sit on the peer-review panels. The STAR project officers have an opportunity to suggest scientific disciplines that should be represented on a panel and provide names of potential reviewers. However, project officers have no role in the selection of panel members.

When a panel has been selected, every member receives abstracts of all the proposals, and three or more panel members are selected to be principal reviewers for each proposal. The principal reviewers receive the entire grant application. For a complex RFA, such as the one issued to establish the multidisciplinary PM centers, 8-10 principal reviewers may be selected. The principal reviewers, who are typically given 3-4 weeks to complete their review, are responsible for reviewing the proposal in detail—ranking it as "excellent," "very good," "good," "fair," or "poor"—and preparing comments based on the following evaluation guidelines (Bryan 2002):

- "**Research Proposal.** Comment on the originality and creativity of the proposed research, the appropriateness and adequacy of the research methods used, and the appropriateness and adequacy of the Quality Assurance Narrative Statement. Also comment on the proposed approach. Is the research approach practical and technically defensible, and can the project be performed within the proposed time period? Could the research potentially contribute to scientific knowledge in the topic area of the solicitation? Is the proposal well prepared with supportive information that is self explanatory and understandable?
- **Investigators.** Comment on the qualifications of the principal investigator(s) and other key personnel, including research training, demonstrated knowledge of pertinent literature, experience, and publication records. Will all key personnel contribute a significant time commitment to the project?
- **Facilities and Equipment.** Comment on the availability and/or adequacy of the facilities and equipment proposed for the project. Identify any deficiencies that may interfere with the successful completion of the research.
- **Responsiveness.** Comment on how well the proposal has responded to the research needs identified. Does the proposal adequately address all of the objectives specified for this topic area?
- **Budget.** Although your evaluation of scientific merit should not be based on budget information, please comment on the appropriateness and/or adequacy of the proposed budget and its implications for the potential success of the proposed research. Input on requested equipment is of particular interest."

When the panel meets, the science review administrator responsible for organizing the panel serves as chair. The STAR project officer is allowed to attend the panel meetings to gain insight into the strengths and weaknesses of the different proposals but is not allowed to influence the panel's deliberations. Panel meetings may continue for 2-3 days. Every proposal that has received at least one "excellent" or two "very good" rankings from the principal reviewers is brought up for review and discussion by the entire panel. Any panel member can request that any other proposal also be discussed.

The panel members discuss the strengths and weaknesses of each proposal until all the members feel qualified to assign it a ranking. The members often agree unanimously to assign it to one of the five categories; when there is no unanimous agreement, an average panel ranking is calculated. One of the principal reviewers of each proposal, designated the rapporteur, then prepares a written summary of the written and oral review comments of all the panel members. The panel summary is signed by all the principal reviewers.

When all the proposals have been addressed, the rankings are reviewed to determine whether there is a consensus that the ranking of any of the individual proposals should be adjusted up or down. Only the proposals that have been ranked in the top two categories—"excellent" and "very good"—are given further consideration. The EPA panel coordinator is responsible for sending a declination letter to each applicant whose proposal did not fall in the top two categories with a copy of the panel summary regarding the proposal.

If an RFA is being jointly funded, the cooperating agency or organization may take on the responsibility of conducting the scientific peer-review process. If EPA is responsible for peer review, the cooperating institution can participate in it under the same restrictions as the STAR project officer and at the conclusion of the process may identify the particular proposals it wishes to fund.

During FY 2001, 24 panels involving a total of 350-400 reviewers were established to review 700-odd grant applications.

For proposals eligible for EPA funding, the next step is a relevance review. This review is intended to ensure that funded projects have high relevance to the agency's mission and program needs. The review is conducted by the STAR project officer, and the procedures are not as formally established as those of scientific peer review. The process involves identifying representatives of the other units of ORD, of the program offices that have an interest in the RFA topic, and of the regional offices to form a relevance-review panel. Those representatives are also subjected to conflict-of-interest review.

The relevance-review panel members are usually supplied with abstracts of all the proposals ranked excellent or very good by the peer-review panel, the summary of comments on these proposals from the peer-review panel, and a list of evaluation criteria. They are typically given 2-3 weeks for their review, which includes preparing written comments and scoring. However, in some cases, they may be given an entire proposal or the detailed reviewer comments, and the time available for review may be adjusted to respond to workloads and work schedules. For both scientific peer review and relevance review, the evaluation criteria are used as a guide, and no specific weighting is given to the individual criteria.

The relevance-review panel meets, typically with some members participating by telephone; discusses each of the proposals; and, by consensus, ranks them from greatest potential value to the agency down to least potential value. The project officer prepares a summary of the relevance-review evaluations and comments and presents it with his or her recommendations to the director of NCER for final funding decisions.

Award Process

At this stage, some flexibility is built into the system. If an unusually high number of excellent, highly relevant proposals have been received, additional funds may be made available to support research conducted under a particular RFA. If few worthy proposals are received, available funding may be decreased. The director of NCER may also modify the ranking of the proposals, although the director usually accepts the ranking presented.[9] The project officer and director may also agree on modifications that the project officer suggests the principal investigators make to specific proposals that have been accepted for funding. The modifications

---

[9]Occasionally, a grant that is not in the highest-priority group from the relevance-review panel is selected because it received an "excellent" rating from the scientific peer-review panel, an apparently parochial ranking has skewed the distribution of the priorities, it will provide better program balance with the rest of the ORD research proposal, a principal investigator on a higher-ranked proposal has not performed well on past STAR grants, or there is a substantial difference of opinion between ORD participants and participants from the rest of the agency (Preuss 2002b).

would be in response to comments made by the review panels or in an effort to make the proposals more responsive to the needs of the agency. Table 2-5 shows the success rate—15%—of proposals submitted in response to RFAs issued in 2001.[10] The success rates in 1999 and 2000 were somewhat lower—11% and 13%, respectively. The success rate in FY 2001 was slightly higher—17% as opposed to 15%—if the RFAs for which ratings information was not available are included. Although the program typically funds a high proportion of the proposals rated excellent by the peer-review panels, it is able to support only about 60% of the proposals rated as excellent or very good.

The project officer may then review the selected proposals to identify budget or other modifications to recommend. The recommendations and a summary of the peer-review panel's comments are sent to the principal investigator, who has to prepare a written response indicating, for each comment or suggestion, how the work and budget will be modified or why no modification is necessary. The interactions between the principal investigator and the project officer at this stage are considered to provide the same advantages as the preproposal discussions used by some other funding organizations without creating concerns about potential conflicts of interest, because funding decisions have already been made. The principal investigator submits a response, a modified proposal and budget, and all the certifications and other submissions required under federal rules, for final processing. The final grant package is reviewed by grant specialists in NCER to ensure that it is complete and is expected to proceed smoothly through the administrative grant-making process before being forwarded to the EPA grants office for funding.

The entire research-grant process takes 1-2 years from the initial announcement of the RFA to the time the grants are actually awarded. The time required for the various steps is indicated in Table 2-6. Much of the time is taken up in responding to administrative requirements; for instance, because the members of a peer-review panel are compensated for the time they spend in reviewing proposals and attending the panel meeting, the agency has to enter into a separate contract with each panel member. NCER

---

[10]Table 2-5 includes only those RFAs for which EPA was able to provide information on how the different proposals were rated by the peer-review panel. EPA was not able to provide such information for several other proposals issued in 2001. When these additional proposals were included in Table 2-5, the overall success rate increases from 15% to 17%.

TABLE 2-5 Grant-Application Success in 2001

Topic	Applications Ranked Excellent		Applications Ranked Very Good		All Applications		
	No.	No. Funded	No.	No. Funded	No.	No. Funded	Success Rate
Air	3	2	12	9	53	11	21%
Water and watersheds[a]	3	3	5	5	26	8	31%
Human health[a]	4	4	20	10	80	14	18%
Ecosystems	0	0	10	3	36	3	8%
Technology	11	10	32	23	185	33	18%
Socioeconomic[a]	3	3	21	14	121	17	14%
Other	7	7	52	18	242	25	10%
Total	**31**	**29**	**152**	**82**	**743**	**111**	**15%**
Success Rate	**94%**		**54%**		**15%**		

[a]Rankings were not available for all applications in this topic, and data include only awards whose rankings were available.
Source: Adapted from J. Puzak, EPA, Washington, D.C., unpublished material, December 12, 2002.

TABLE 2-6 Time Requirements for Processing Research Grants

Processing Step	Time Required	Comments
Preparation and approval of RFA	1-3 months	Can take longer if other agencies involved
Announcement open (time from announcement to issue of deadline for proposals)	3-5 months	Larger and more-complex proposals (such as research centers) are given more time
Peer-review panel	3-4 months	Have to enter into individual contract with each panel member
Relevance review	4-6 weeks	Depends on conflicting demands, such as budget preparation, on EPA staff members involved in review
NCER director's approval	2-3 weeks	
Researchers prepare final proposals and required documentation	1-3 months	
Final review and signoff on grant packages in ORD	1-2 months	Could be shortened with addition of administrative staff
Processing by EPA grants administration	1-2 months	

Source: B. Levinson, EPA, personal commun., November 6, 2002.

is hiring additional administrative staff to deal with some of the administrative requirements. That may not noticeably speed up the grant-making process, but it will at least reduce the project officers' workload and allow them to pay more attention to the substance of the research efforts, to monitoring research progress, and to communicating research results to potential users.

One important reason for undertaking such a careful final review of a proposal, with its sometimes extensive interaction with the principal investigator, is that almost all STAR grants for individual researchers are fully funded when they are awarded. (Research centers, however, are funded on an annual basis as long as sufficient funding is included in the agency's appropriations and the centers perform adequately.) Even when the agency does not initially fund a grant fully, it makes a commitment to the grantee to provide complete funding as long as Congress appropriates sufficient money and the grantee satisfies federal administrative requirements and submits the required annual progress reports. Commitments that have not been fully funded are given first priority for funding in later budget years (J. Puzak, EPA, Washington, D.C., personal commun., November 6, 2002).

Implementation and Evaluation

During the 3-5 years of a grant period (many grantees also receive, at their request, a 1-year, no-cost extension to the grant period), the STAR project officer is expected to monitor grant performance, including the submission of the annual progress reports and the grantee's compliance with federal requirements, such as the OMB data-quality guidelines. Project officers also attempt to visit all the research centers and institutions that receive large individual grants to check on research progress.

A major mechanism for evaluating research in progress is the scheduling of progress-review workshops, which are focused on specific topics, as determined by the project officer, and bring together all the STAR researchers and ORD staff working on the topic in question to discuss their progress and issues that have appeared in the course of the research. The key EPA-funded researchers and representatives of ORD laboratories doing associated research are expected to attend these meetings. Representatives of other EPA offices and federal agencies cooperating in the research effort are also invited, and the meetings are open to the public.

In addition to providing an opportunity for peer review of research in progress, the meetings allow researchers to interact with one another. Although the greatest value of the workshops is probably the information exchange that occurs, EPA may also make the workshop proceedings available to the public. EPA reports that progress-review "workshops have been held in every program area, although there is not an annual workshop for every program area" (Preuss 2002b).

With those monitoring mechanisms, the agency believes that it tracks grantees "more closely than other agencies" (Preuss 2002b).

Efforts to coordinate STAR-sponsored research with the other units of ORD and with the program offices continue while the research is being conducted. If strong coordination is desired, the funding mechanism can be made a "cooperative agreement," which allows EPA staff to work cooperatively with the researchers, rather than a grant, which requires the researchers to work independently. The progress-review meetings are another mechanism for coordination. At a minimum, EPA staff and the public have access to the annual progress reports, which are posted on the NCER Web site for STAR research grants, and the program and regional offices can request the preparation of STAR "research capsule reports" (described below).

Dissemination of Results

The other activity that occurs after grants are awarded is the dissemination of information about the research efforts to potential users and the public. Many research programs pay little attention to dissemination, relying primarily on the researchers themselves, using such normal academic communication channels as conferences and journals, to spread the word. However, efforts through the STAR program are more aggressive. The office provides information concerning each grant that it awards on its Web site shortly after the award is made. It then requires grantees to prepare annual progress reports, and these too are made available to the public through the STAR Web site. A recently improved search engine allows users to quickly obtain a list of all STAR projects and reports dealing with a particular subject.

In addition to the individual project reports, the STAR program prepares and publishes the following reports on its Web site:

- STAR reports, typically 4-6 pages long, which provide summary descriptions of research in progress on selected research topics for the general public. As of November 2002, STAR had released 10 of these reports (EPA 2002c).
- STAR research capsules, prepared at the request of EPA program and regional offices, which provide brief summaries of all the individual research projects that STAR is supporting on specific scientific issues. As of November 2002, STAR had released 18 of these reports (EPA 2002d).
- Progress-review workshop proceedings, which contain the presentations made at selected progress-review workshops. As of November 2002, NCER had released nine of these proceedings, although it had sponsored many more workshops than that (EPA 2002e).
- State-of-the-science reports, which will summarize all the current scientific information related to selected environmental issues, regardless of who sponsored the research that provided the information. These reports are being prepared by contractors; none had been released by November 2002.

In addition to making that information available to those who have sufficient interest to visit and search the STAR Web site, STAR program officers apparently spend a substantial amount of time, including weekly conference calls, communicating research progress to the EPA program and other ORD offices (EPA/BOSC 2002). Communication to other potential user groups is more difficult, but the STAR program has experimented with some unusually aggressive mechanisms. In the mid-Atlantic region, for instance, a special pilot project was established to communicate STAR efforts to state and local agencies and public-interest groups and to determine the apparent relevance of the STAR research to the missions of the agencies and groups (Bradley 2002). In November 2002, STAR and EPA Region 1 (New England) sponsored a workshop at which STAR grantees in the region discussed their research with a similarly diverse group of interested parties. The workshop was well attended, and the participants strongly supported the program's effort to communicate its research findings more quickly and efficiently (E. Abt, NRC staff, Washington, D.C., personal commun., Dec. 11, 2002). Apparently, the workshop was considered so successful that similar efforts are being considered in other regions.

In spite of those efforts, the effective dissemination of research results to the diverse potentially interested audiences remains a challenge and is one of the issues highlighted in several reviews of the STAR program (for example, EPA/BOSC 2002).

FELLOWSHIPS

The process for awarding STAR fellowships is similar to but simpler than the process of awarding research grants. A notice indicating the availability of the grants, the eligibility criteria, the submission requirements, and the deadline for submissions is published in the *Federal Register* and on the STAR Web site, and copies are sent to graduate schools that have programs in environmental sciences and to individuals and organizations that have requested notification. The fellowship announcement is generally posted in mid-August and remains open for about 90 days, closing in mid-November. Potential applicants can sign up on the Web site to receive e-mail notifications about fellowships.

The fellowship applications are submitted to the Peer Review Division of NCER, which conducts the peer review for the applications. That division first reviews the applications to ensure that they are complete and satisfy the eligibility requirements. It then establishes a fellowship review panel composed of academics. The panel members are subject to similar conflict-of-interest requirements as members of the peer-review panels for grants.

Each fellowship application is assigned to three panel members designated as principal reviewers. They are responsible for reviewing, ranking, and preparing written comments on each of the applications they are assigned. The panelists receive the applications 4 weeks before the peer review meeting and assign each application to one of five categories—"excellent," "very good," "good," "fair," or "poor" on the basis of the following criteria (J. Gentry, EPA, Washington, D.C., personal commun., August 9, 2002):

- "**Goals and objectives** (all applicants). Comment on the seriousness of the applicant's dedication to the stated career goals and objectives. Comment on the student's organizational, analytical, and written skills.
- **Entering master's student** (applicants who at the time of submission are applying for or enrolled in a master's program and have completed less than 1 year toward this degree). Comment on the strength of the applicant's planned course of study and probability of success of any proposed project.
- **Entering doctoral student without another graduate degree** (applicants who at the time of submission are applying for or are enrolled in a doctoral program, have completed less than 1 year toward this degree, and have no other graduate or other professional degree [MS, DVM, or JD]). Comment on the strength of the applicant's planned course of study and probability of success of any proposed project.
- **Beginning doctoral student with another graduate degree** (applicants who at the time of submission are applying for or enrolled in a doctoral program, have completed less than 1 year toward this degree, but have completed another graduate or professional degree [such as MS, DVM, MD]). Comment on how the applicant's proposed doctoral program builds on his/her previous education or research projects. Why and how will any proposed research project advance the applicant's academic or career goals?
- **Continuing doctoral student** (applicants who at the time of submission are enrolled in a doctoral program and have completed more than 1 year, but less than 4 years toward this degree). Comment on the applicant's research project as to its technical and social application, potential for success, and benefits expected."

During the panel meeting, only applications that have been rated excellent or very good by the principal reviewers typically are discussed, although a principal reviewer can ask that a particular application be discussed even if it has not received a sufficiently high ranking by all three principal reviewers. After the panel reaches agreement on the ranking of an application, a panel summary is prepared that contains a summary of the review comments and panel ranking; the summary is provided to the applicant.

Typically, more applications are ranked excellent by the peer-review panel than EPA can afford to fund. The final decision about which of the "excellent" applications receive funding is made by the NCER staff according to such criteria as achieving a balance of fellowships among universities, filling identified shortfalls in particular disciplines, achieving a rough proportion among disciplines between the number of "excellent" applications and the number of fellowships awarded, and emphasizing applications in disciplines that EPA considers particularly important to fulfilling its science mission. The emphasis is on disciplines rather than on the specific research that applicants intend to undertake and whether it is relevant to EPA's mission. The entire review process is completed by March. Applicants are notified and first-year funding provided to successful applicants in June.

The fellowship awards are for 2 years (for master's candidates) or 3 years (for Ph.D. candidates). EPA issues each successful applicant a letter committing the agency to the full term of funding contingent on the applicant's remaining a student in good standing as demonstrated by an annual report provided by the student's academic institution and academic adviser. Fellowships may be terminated if the EPA project officer determines that the fellow is not performing up to the standards of the program.

PREVIOUS EVALUATIONS OF THE PROGRAM

Given its relatively short life, the STAR program has been subject to an unusual number of reviews and evaluations by advisory committees and other internal and external groups (Table 2-7). The present review by a committee of the National Research Council is the most recent, and the committee has been informed that the EPA Office of the Inspector General is in the process of, or at least considering, undertaking another review (Harris 2002). NCER invited some of the reviews (such as this one) to obtain independent evaluations of its success in implementing an effective research program and suggestions on how the program could be improved. Others have been imposed on the organization.

TABLE 2-7 Summary of Reviews of the STAR Program

Year	Author	Title of Report
1998	Board of Scientific Counselors (BOSC) of EPA's ORD	*Program Review of the National Center for Environmental Research and Quality Assurance (NCERQA). Final Report of the Ad Hoc Subcommittee on the Review of NCERQA* (EPA/BOSC 1998).
2000	EPA's Science Advisory Board and ORD's BOSC	*A Joint SAB/BOSC Report: Review of the Science to Achieve Results (STAR) Program* (EPA/SAB/BOSC 2000).
2000	General Accounting Office (U.S. Congress)	*Environmental Research: STAR Grants Focus on Agency Priorities, but Management Enhancements are Possible* (GAO 2000).
2001	EPA's Science Advisory Board	*The Science to Achieve Results (STAR) Water and Watersheds Grants Program: An EPA Science Advisory Board Review* (EPA/SAB 2001b).
2001	EPA and National Science Foundation	*Evaluation Report: A Decision Making and Valuation for Environmental Policy Interim Assessment* (EPA/NSF 2000).
2002	EPA's Science Advisory Board	*Interim Review of the Particulate Matter (PM) Research Centers of USEPA: An SAB Report* (EPA/SAB 2002a).
2002	ORD's BOSC	*Program Review of NCER* (EPA/BOSC 2002).
?	EPA's Office of the Inspector General	EPA's Office of the Inspector General is considering conducting another review of the STAR program (Harris 2002).

The recommendations from those reviews and EPA's initial responses to them are summarized in Appendix B. Almost all the reviews have been generally supportive of the program and have focused on how it might be even more effective. The STAR program appears to have implemented many of the recommendations, and these recommendations have contributed to the modifications of the program described in this chapter.

Although the committee believes that objective, independent reviews of government programs can have substantial benefits, it also recognizes that such reviews can be expensive in the resources devoted to them and in the disruption that they can cause in the organization being reviewed. At the very least, anyone considering another review should carefully consider the results of previous efforts to ensure that the new review will truly be valuable.

CONCLUSIONS AND RECOMMENDATIONS

- The STAR program is a crucial element of EPA's research efforts. It allows the agency to fill information gaps that are not addressed completely by its intramural program and to respond to new issues that EPA laboratories are not able to address. In addition to those primary purposes of the program, it provides the agency access to independent information, analyses, and perspectives. It helps to maintain environmental research and analysis capabilities in many of the nation's academic and nonprofit research institutions. Finally, the program provides for the education of future leaders in environmental science and engineering. For these reasons, the STAR program should continue to be an important part of EPA's research program.
- As the STAR program has evolved, it has developed a grant-award process that in many ways exceeds those in place at other organizations that have extramural research programs. The agency has an aggressive planning process to identify the specific research that should be supported. The scientific peer-review process has been well established, and the proper mechanisms are in place to avoid conflicts of interest and to ensure independent reviews. The agency also puts an unusual amount of effort into preparing research solicitations and funding projects that have high relevance to its mission and program needs.
- As the STAR program has developed, it has been able to induce other agencies with similar interests to enter into partnerships and provide supplementary funds. STAR should continue to partner with other govern-

ment and nongovernment organizations to support research of mutual interest and of relevance to EPA's mission, explore innovative approaches for carrying out this research, and sponsor a diverse portfolio of research that alerts the agency to emerging issues and provides independent analyses of issues that the agency is currently addressing.

- When projects are under way, the STAR program actively monitors their progress and coordinates the efforts of the independent researchers with one another as appropriate and with the conduct of related research by EPA staff.
- The STAR program has begun to emphasize communication of research results to potential users and has taken some remarkably aggressive steps to promote it.

REFERENCES

Bradley, P. 2002. Mid-Atlantic Integrated Assessment (MAIA) Pilot Program Overview. Presentation at the First Meeting on Review of EPA's Research Grants Program, March 19, 2002, Washington, DC.

Bryan, E. 2002. Peer Review Process for EPA's National Center for Environmental Research's Science to Achieve Results Program. Presentation at the Third Meeting on Review of EPA's Research Grants Program, June 6, 2002, Washington, DC.

College Board. 2001. Trends in College Pricing 2001. College Board Publications, New York.

EPA (U.S. Environmental Protection Agency). 1996a. Report to Congress. The Science to Achieve Results (STAR) Program. EPA/600/R-96/064. Office of Research and Development, U.S. Environmental Protection Agency, Washington, DC [Online]. Available: http://es.epa.gov/ncer/publications/archive/reportcong.html [accessed Jan. 13, 2003].

EPA (U.S. Environmental Protection Agency). 1996b. Research Opportunities. U.S. Environmental Protection Agency Announces the Availability of 1996 Grants for Research on Endocrine Disruptors Role of Interindividual Variation in Human Susceptibility to Cancer Risk- Based Decisions for Contaminated Sediments. National Center for Environmental Research, Office of Research and Development, U.S. Environmental Protection Agency [Online]. Available: http://es.epa.gov/ncer/rfa/archive/grants/96/rfa2.html [accessed Jan. 13, 2003].

EPA (U.S. Environmental Protection Agency). 1997. Ecosystem Indicators. Research Opportunities. National Center for Environmental Research, Office of Research and Development, U.S. Environmental Protection Agency [Online]. Available: http://es.epa.gov/ncer/rfa/archive/grants/97/ecosystem.html [accessed Jan. 13, 2003].

EPA (U.S. Environmental Protection Agency). 1998. Ecological Research Strategy. EPA/600/R-98/086. Office of Research and Development, U.S. Environmental Protection Agency, Washington, DC [Online]. Available: http://www.epa.gov/ordntrnt/ORD/WebPubs/final/eco.pdf [accessed Jan. 13, 2003].
EPA (U.S. Environmental Protection Agency). 1999. 1999 STAR Grants for Research: Ecological Indicators. Research Opportunities. National Center for Environmental Research, Office of Research and Development, U.S. Environmental Protection Agency [Online]. Available: http://es.epa.gov/ncer/rfa/archive/grants/99/batch.html [accessed Jan. 13, 2003].
EPA (U.S. Environmental Protection Agency). 2001. Office of Research and Development Strategic Plan. Office of Research and Development, U.S. Environmental Protection Agency [Online]. Available: http://www.epa.gov/ospinter/strtplan/documents/final.pdf [accessed Jan. 13 2003].
EPA (U.S. Environmental Protection Agency). 2002a. Welcome to EPA's National Center for Environmental Research Homepage. National Center for Environmental Research, Office of Research and Development, U.S. Environmental Protection Agency [Online]. Available: http://es.epa.gov/ncer/about/ [accessed Jan. 13, 2003].
EPA (U.S. Environmental Protection Agency). 2002b. Airborne Particulate Matter (PM) Research Centers. National Center for Environmental Research, Office of Research and Development, U.S. Environmental Protection Agency [Online]. Available: http://es.epa.gov/ncer/centers/airpm/ [accessed Jan. 13, 2003].
EPA (U.S. Environmental Protection Agency). 2002c. STAR Reports. National Center for Environmental Research, Office of Research and Development, U.S. Environmental Protection Agency [Online]. Available: http://es.epa.gov/ncer/publications/starreport/index.html [accessed Jan. 13, 2003].
EPA (U.S. Environmental Protection Agency). 2002d. STAR Research Capsules by Topic. National Center for Environmental Research, Office of Research and Development, U.S. Environmental Protection Agency [Online]. Available: http://es.epa.gov/ncer/publications/topical/topic.html [accessed Jan. 13, 2003].
EPA (U.S. Environmental Protection Agency). 2002e. Progress-Review Workshop Proceedings. National Center for Environmental Research, Office of Research and Development, U.S. Environmental Protection Agency [Online]. Available: http://es.epa.gov/ncer/publications/workshop/ [accessed Jan. 13, 2003].
EPA (U.S. Environmental Protection Agency). 2002f. Fall 2002 Science to Achieve Results Fellowships for Graduate Environmental Study. 2002 RFA. National Center for Environmental Research, Officer of Research and Development, U.S. Environmental Protection Agency.
EPA (U.S. Environmental Protection Agency). 2003a. ORD Lab & Office Locations. Office of Research and Development, U.S. Environmental Protection Agency [Online]. Available: http://www.epa.gov/ord/htm/map.htm [accessed Jan. 13, 2003].

EPA (U.S. Environmental Protection Agency). 2003b. Research Opportunities - Environmental Research Grant Announcements. National Center for Environmental Research, Office of Research and Development, U.S. Environmental Protection Agency [Online]. Available: http://es.epa.gov/ncer/rfa/ [accessed Jan. 13, 2003].

EPA (U.S. Environmental Protection Agency). 2003c. STAR Fellowships. National Center for Environmental Research, Office of Research and Development, U.S. Environmental Protection Agency [Online]. Available: http://es.epa.gov/ncer/fellow/ [accessed Jan. 13, 2003].

EPA/BOSC (U.S. Environmental Protection Agency Board of Scientific Counselors). 1998. Program Review of the National Center for Environmental Research and Quality Assurance (NCERQA). Final Report of the Ad Hoc Subcommittee on the Review of NCERQA. Board of Scientific Counselors, Office of Research and Development, U.S. Environmental Protection Agency, Washington, DC. April 30, 1998.

EPA/BOSC (U.S. Environmental Protection Agency Board of Scientific Counselors). 2002. Program Review of NCER. Board of Scientific Counselors, Office of Research and Development, U.S. Environmental Protection Agency, Washington, DC. October 25, 2002.

EPA/NSF (U.S. Environmental Protection Agency and National Science Foundation). 2000. Interim Assessment for the Decision Making and Valuation for Environmental Policy Grants Program. Final Report. Prepared for National Science Foundation and U.S. Environmental Protection Agency, by Aspen Systems Corporation. April 17, 2000.

EPA/SAB (U.S. Environmental Protection Agency Science Advisory Board). 2001a. Water Quality and Pollution Prevention Multiyear Plans: An SAB Review. A Review by the Research Strategies Advisory Committee (RSAC) of the EPA Science Advisory Board (SAB). EPA-SAB-RSAC-02-003. Science Advisory Board, U.S. Environmental Protection Agency, Washington, DC [Online]. Available: http://www.epa.gov/science1/fiscal02.htm [accessed Jan. 13, 2003].

EPA/SAB (U.S. Environmental Protection Agency Science Advisory Board). 2001b. The Science to Achieve Results (STAR) Water and Watersheds Grants Program: An EPA Science Advisory Board Review. A Review by the Ecological Processes and Effects Committee (EPEC) of the EPA Science Advisory Board. EPA-SAB-EPEC-02-001. Science Advisory Board, U.S. Environmental Protection Agency, Washington, DC [Online]. Available: http://www.epa.gov/science1/fiscal02.htm [accessed Jan. 13, 2003].

EPA/SAB (U.S. Environmental Protection Agency Science Advisory Board). 2002. Interim Review of the Particulate Matter (PM) Research Centers of the USEPA: An EPA Science Advisory Report. A Review by the PM Research Centers Interim Review Panel of the Executive Committee of the U.S. EPA Science Advisory Board (SAB). EPA-SAB-EC-02-008. Science Advisory Board, U.S. Environmental Protection Agency, Washington, DC. May 2002

[Online]. Available: http://www.epa.gov/science1/fiscal02.htm [accessed Jan. 13, 2003].

EPA/SAB/BOSC (U.S. Environmental Protection Agency Science Advisory Board and Board of Scientific Counselors). 2000. A Joint SAB/BOSC Report: Review of the Science to Achieve Results (STAR) Program. EPA-SAB-EC-00-008. Science Advisory Board, Board of Scientific Counselors, U.S. Environmental Protection Agency. [Online]. Available: http://www.epa.gov/sab/pdf/ec0008.pdf [accessed Jan. 13, 2003].

GAO (U.S. General Accounting Office). 2000. Environmental Research: STAR Grants Focus on Agency Priorities, But Management Enhancements Are Possible: Report to the Chairman, Subcommittee on VA, HUD, and Independent Agencies, Committee on Appropriations, House of Representatives. GAO/RCED-00-170/B-142370. U.S. General Accounting Office, Washington, DC.

Harris, J.K. 2002. Use of Science to Achieve Results (STAR) Grants in Achieving EPA's Strategic Goals. Memorandum to H.L. Longest II, Acting Assistant Administrator for Research and Development, from J.K. Harris, Director for Cross-Media Issues, Office of Program Evaluation, Office of the Inspector General, Washington, DC. March 13, 2002.

Hogue, C. 2002. Graduate program cut from budget. Chem. Eng. News 80(17): 22.

Johnson, J. 1996. Rebuilding EPA science. Environ. Sci. Technol. 30(11):492A-497A.

NCSE (National Council for Science and the Environment). 2003. Description of the EPA STAR Fellowship Program. National Council for Science and the Environment, Washington, DC [Online]. Available: http://www.cnie.org/NCSE/SciencePolicy/? [accessed Jan. 13, 2003].

NRC (National Research Council). 1998. Research Priorities for Airborne Particulate Matter. 1. Immediate Priorities and a Long-Range Research Portfolio. Washington, DC: National Academy Press.

Preuss, P.W. 2002a. National Center for Environmental Research, History, Goals, and Operation of the STAR Program. Presentation at the First Meeting on Review of EPA's Research Grant Program, March 18, 2002, Washington, DC.

Preuss, P.W. 2002b. Response to NAS Committee Queries. Presentation at the Third Meeting on Review of EPA's Research Grant Program, June 6, 2002, Washington, DC.

3

Competitive Grant Programs in Other Federal Agencies

To assess the Environmental Protection Agency (EPA) Science To Achieve Results (STAR) program's progress in establishing a competitive grants program, the committee examines the procedures put into place by other federal agencies. The committee was tasked specifically with addressing the STAR program in relation to other research grants programs. The agencies selected for review were ones that have partnered with STAR in supporting joint research. These agencies represent a broad spectrum of basic and applied research agencies (See Box 3-1).

The National Science Foundation (NSF) and National Institutes of Health (NIH) are often thought of as organizations whose principal mission is to support basic research and the progress of science itself. However, other agencies that have defined service missions related to national security and public welfare also have a long history of supporting fundamental and exploratory research, which informs their missions and is supported by academic expertise. All those agencies have established competitive, peer-review processes to evaluate and select individual investigator and multi-investigator projects. Depending on the mission and the expected effects of the research, the agencies have differing approaches to reviewing their projects and programs and disseminating the resulting information.

This chapter describes the program-management procedures of the agencies listed in Box 3-1. It addresses the following research-management

BOX 3-1 Federal Agencies and Their Research Grants

Agency	*Research Organization*
National Science Foundation	Environmental Research and Education
National Institutes of Health	National Institute of Environmental Health Sciences
Department of Energy	Office of Biological and Environmental Research Competitive Grants Program
Department of Agriculture	National Research Initiative
National Aeronautics and Space Administration	Biological and Physical Science Research
National Oceanic and Atmospheric Administration	Office of Oceanic and Atmospheric Research

activities: planning, opportunity communication, proposal review process, agency involvement during implementation, project and program evaluation, and dissemination of results. It is important to note that those research activities often overlap and have strong feedback mechanisms as they are carried out by agency program managers. Planning affects project review criteria, project and program evaluation feeds into planning, and so on. Table 3-1 summarizes the procedural aspects of the various research programs that are reviewed here in comparison with that of the STAR program. The committee did not address the issue of cost comparisons to operate the various federal research programs, as this was not part of their charge. Information in this chapter comes from individual discussions with and presentations by agency personnel and from publicly available documentation from the agencies.

THE NATIONAL SCIENCE FOUNDATION

NSF is the principal federal agency with a mission of promoting the progress of science in support of the national welfare and defense (Firth 2002). It is structured to reflect the scientific and engineering disciplines.

TABLE 3-1 Summary of Extramural Grants Programs

Program	Planning	Communication	Proposal Review	Implementation	Evaluation	Dissemination
EPA STAR	Planning process is variable and highly dependent on specific program; programs are conducted within context of multiyear plans and budget plans that are revised at least annually but generally not reviewed externally.	RFA is published in *Federal Register*; program offices post the RFA on Web site and announce at professional meetings; in some cases, program officer sends information via NCER listserv to interested parties.	Peer-review process is conducted by NCER's peer-review division, which handles review of proposals.	POs do not have much ability to influence implementation; there is some opportunity for coordination and integration with researchers; however, this varies from PO to PO.	Evaluations have been done on a program-specific basis and have been conducted by EPA's SAB, BOSC, and GAO.	Methods of dissemination include "synthesis" reports, "state of the science" reports, and annual progress meetings.
NSF	Strategic outcome goals (people, ideas, and tools) set context for investment priorities; priorities are defined by community pressure, disciplinary consensus agendas, professional meeting agendas, review panels and advisory committees, staff input, director's office and White House guidance.	RFAs are broadcast through *Federal Register* and NSF Web site. Competitions for center grants and those associated with interagency partnerships are announced through RFAs. Most proposals to NSF are unsolicited and investigator-initiated.	Volunteers participate in NSF review process through combination of panel review, letter review, and site visits as appropriate; POs play important role in process.	Grant mechanism provides for arm's-length relationship during course of relationships among investigator, institution, and NSF; PO may choose to visit investigator during course of grant.	Projects are evaluated by publications in refereed journals, presentations at professional meetings, involvement of students, advisory committees, and committees of visitors.	Results are disseminated through journals, professional meetings, and regular meetings of grantees and contractors.

NIEHS	Annually, NIEHS divisions undertake planning process to identify areas of emphasis for coming fiscal year; areas of emphasis may be reflected in changes to general program announcement that NIEHS releases; the institute may also decide to release specific RFA for one-time competition in well-defined scientific field.	Program announcements and RFAs are released through standard federal publications and on institute's Web site. Majority of 40,000 proposals received by NIH are unsolicited.	Unsolicited proposals and responses to general program announcements are reviewed through CSR. Proposals that respond to RFAs are reviewed within specific institute.	Grant mechanisms provide for arm's-length relationship between investigator and institute; institutes use cooperative agreements and contracts that provide institute staff with considerably more influence; institute staff may choose to visit investigator during course of project with mutual benefit.	Projects are evaluated by publications in refereed journals, presentations at professional meetings, involvement of students, institute's National Advisory Council, and its Board of Scientific Counselors.	Results are disseminated through journals and professional meetings; NIH planning meetings also bring together researchers to report on their work and to provide interactions that may result in future proposals.
DOE BER	Factors considered in planning: national need, DOE mission need, opportunities consistent with DOE capabilities and mission, critical scientific gaps, and	Continuing program announcement for entire office of science is published once a year through *Federal Register* and SC Web site; BER-	Mechanisms for peer review: individual mail reviews, ad hoc committees, or standing committees as	Grant mechanism provide for arm's-length relationship between official and investigator;	BER conducts retrospective reviews of its research in both grant programs and laboratory research; these reviews are conducted by its	Publication and meeting presentations are significant measures of individual projects.

(Continued)

TABLE 3-1 *Continued*

Program	Planning	Communication	Proposal Review	Implementation	Evaluation	Dissemination
DOE BER (Cont.)	congressional mandate; mechanisms used to define priorities: FACA advisory committee, commissioned scientific assessments, scientific workshops, program review findings, and recommendations and staff interactions with scientific community.	specific solicitations are publicized in trade and professional journals and in *Federal Register* and SC Web site.	appropriate or required by legislation; each application is reviewed by at least three qualified reviewers and conflict-of-interest issues are overseen by program manager.	DOE program manager visits investigators at both universities and national laboratories on regular basis; BER programs also bring university and laboratory investigators together at regular meetings.	FACA committees and in some cases the JASON organization and NRC boards and committees.	Participation by principal investigators in international or interagency coordination committees is also a measure of project results.
NRI	Opportunities are identified through proposal submissions of research communities and response and internal discussions of NRI peer-review panels; planning process includes staff participation in interagency coordination	NRI program description is made available in printed form, on the Internet, in the *Federal Register*, and at scientific meetings and conferences.	Proposals are assigned to three panelists and four to six ad hoc reviewers; panel meeting is held to discuss and rank proposals from 1 to 6; awards are based on panel's ranking; NRI staff finalize budget but	Although initial terms and conditions govern performance of grant, there is no official mechanism for NRI program managers to direct principal investigators	NRC has provided reviews of NRI in 1994 and 2000; initiative also evaluates itself against CREES's GPRA outcomes by providing appropriate descriptions of project impacts on those outcomes; NRI Board of Directors	Dissemination activities include publication in refereed journals, presentations, scientific conferences, graduate students, NRI staff

	committees, regular staff meetings with commodity groups and agricultural coalition organizations, and congressional input through NRI budget process.		do not adjust individual project budgets.	during term of grant.	provides oversight of NRI policy retrospectively and prospectively.	presentations of results at meetings with various USDA stakeholders, and preparation and distribution of one-page summaries of particular results.
NASA	The following contribute to NASA strategic plan: research divisions, standing committees of NASA Advisory Council that undertake program reviews and planning activities that affect divisions' plans, congressional direction through budget process, and studies of NRC Space Studies Board and Aeronautics and	NASA research announcements are communicated on NASA Web site, in *Commerce Business Daily*, and *Federal Register*; community involvement is part of communication effort.	Letter and panel reviews are used; also, NASA uses support-service contractor to carry out logistics of peer review; contractor maintains proposal and reviewer databases; program manager suggests reviewers and works with contractor to	Program managers' interaction with grantees is similar to that in other agencies; however, program managers often take more active role by organizing formal and informal meetings,	NAC conducts annual reviews of research divisions to assess their performance against their GPRA measures; NAC standing committees do retrospective reviews of specific elements of divisions; NASA also establishes external ad hoc review committees; NRC has	NASA has statutory mandate and receives specific funding for publicizing its mission and its results; results are disseminated through journals and professional meetings and

(Continued)

TABLE 3-1 *Continued*

Program	Planning	Communication	Proposal Review	Implementation	Evaluation	Dissemination
NASA (Cont.)	Space Engineering Board.		ensure conformance with conflict-of-interest rules.	including professional-society meetings.	reviewed specific NASA programs (for example, through Space Studies Board and Aerospace and Engineering Board).	to public media.
NOAA	Each NOAA unit contributes to strategic plan; OAR receives considerable input from major research universities, environmental managers, and general public via Sea Grant network in each coastal state; input is coordinated by three OAR Assistant Deputies for agency cross-cutting and interdisciplinary research activities.	Research opportunities are communicated in *Federal Register*, *Commerce Business Daily*, direct mailings, and NOAA, the Joint Institutes Program, and Sea Grant Web sites.	Letter reviews and panel reviews are used; reviewers are asked to evaluate proposals on basis of scientific merit, whether proposed work addresses Agency's mission, and mission of division; additional considerations include cost, opportunities for outreach activities, qualifications of investigators and key scientific personnel, and institutional	Strong incentives exist for interaction between program managers to visit grantees, stay abreast of program issues, and ensure prompt input into the research results; Sea Grant program also provides opportunities for interaction between researchers at different	Joint Institute Program, Arctic Research Program, and Office of Global Change Program are all evaluated via regular peer-review visits.	Research is disseminated through workshops, scientific meetings, and journal articles; because Sea Grant College Program's mission includes research, outreach, and educational components, it has developed criteria and benchmarks for "connect-

		research infrastructure support.	institutions on regular basis.		ing with the user."

Abbreviations: BOSC, Board of Scientific Counselors; CREES, Cooperative State Research, Education, and Extension Service; CSR, Center for Scientific Review; DOE BER, Department of Energy Office of Biological and Environmental Research; EPA STAR, Environmental Protection Agency Science To Achieve Results; FACA, Federal Advisory Committee Act; GAO, General Accounting Office; GPRA, Government Performance and Results Act; NAC, NASA Advisory Council; NASA, National Aeronautics and Space Administration; NCER, National Center for Environmental Research; NIEHS, National Institute of Environmental Health Sciences; NIH, National Institutes of Health; NOAA, National Oceanic and Atmospheric Administration; NSF, National Science Foundation; NRC, National Research Council; NRI, National Research Initiative; OAR, Office of Ocean and Atmospheric Research; POs, project officers; RFA, Request for Application; SAB, Scientific Advisory Board; SC, Office of Science; USDA, U.S. Department of Agriculture.

Its FY 2002 budget was about $4.5 billion, of which about $825 million supported the Working Group on Environmental Research and Education (ERE 2003).

Planning

With respect to priority-setting and planning, NSF has three strategic outcome goals (Firth 2002):

- **People.** Developing a diverse, internationally competitive and globally engaged workforce of scientists, engineers, and well-prepared citizens.
- **Ideas.** Enabling "discovery across the frontier of science and engineering connected to learning, innovation, and service to society."
- **Tools.** Providing broadly accessible, state-of-the-art and shared research and education tools.

Those goals set the context for research priorities that are defined in various ways through community pressure, disciplinary consensus agendas, professional-meeting agendas, review panels and advisory committees, staff input, and Director's Office and White House guidance (for example, in interagency budget initiatives).

Opportunity Communication

Requests for applications (RFAs) are used and broadcast through the *Federal Register* and the NSF Web site. Competitions for center grants are also announced through RFAs, as are competitions associated with interagency partnerships. However, most proposals to NSF are still unsolicited and investigator-initiated. As priorities change in the planning process, new competitions may be initiated, new emphases may be established, and other emphases may be terminated.

Proposal Review

NSF reviews about 30,000 proposals per year and makes 10,000 new awards to 2,000 colleges, universities, elementary and secondary schools, nonprofit institutions, and small businesses. Around 50,000 scientists and engineers participate as volunteers in the NSF review process through a combination of panel reviews, letter reviews, and site visits. Expenses are covered, and a small honorarium is provided, but reviewers are not compensated for their time.

The role of the program officer (PO) is important in this process. The POs are subject-matter experts and know the communities from which reviewers should be chosen to provide an effective and fair review. POs oversee the NSF conflict-of-interest and confidentiality policy, but how well the process operates depends ultimately on the honesty of the reviewers.

Implementation

The NSF grant mechanism provides for an arm's-length relationship between the investigator, the institution, and NSF. The PO may choose to visit the investigator during the course of the grant, and such interactions may be mutually beneficial to the PO and the researcher. The investigator gains the broader perspective of the PO as to progress in the broad community; the PO gains additional and prompt information that may affect future research-program directions.

Evaluation

Individual projects are evaluated on the basis of publications in refereed journals, presentations at professional meetings, and participation of students in the research. The success of a program or division in supporting a particular field of research or in developing a new multidisciplinary research field may be evaluated by advisory committees or by a more focused committee of visitors. Such committees assess not only the progress of the projects and the overall program but also the success of the POs in reviewer selection and process efficiency.

Dissemination of Results

In general, results of NSF-supported research are disseminated through the normal mechanisms of the research community: journals and professional meetings. However, for interdisciplinary and interagency programs, regular meetings of grantees and contractors may also be called by project officers to ensure effective communication and the development of useful networks and collaborations.

NATIONAL INSTITUTE OF ENVIRONMENTAL HEALTH SCIENCES IN NIH

The National Institute of Environmental Health Sciences (NIEHS) is one of 20 institutes and seven centers of NIH. Each institute and center is organized similarly, as indicated in Figure 3-1. Each institute supports intramural and extramural research. Across the institutes in 2001, nearly 90% of the $18.6 billion budget supported extramural research. The institutes use a variety of procurement mechanisms: grants, cooperative agreements, and contracts.

NIEHS's mission is to reduce the burden of human illness and disease from environmental causes. It examines the character and interrelated effects of environmental exposures, individual susceptibility, age, and length of exposure. NIEHS supports research that ranges from fundamental research in molecular toxicology to the study of disease effects and includes prevention and economic consequences (Thompson 2002). NIEHS's FY 2001 budget was about $1.47 billion.

NIEHS has collaborated with STAR since 1995 in research on endocrine disruptors and complex chemical mixtures. NIEHS and EPA are cosponsors of the 12 centers for children's environmental health.

Planning

NIEHS has identified several topics of interest for long-term investment:

- Genomic imprinting and environmental disease susceptibility.
- Fetal origins of adult diseases.
- Oxidative stress and dietary modulation.
- Human health effects of complex mixtures.
- Environmental factors in diseases of concern in women and men.
- Ethical, legal, and social implications of human genetics and genomics research.

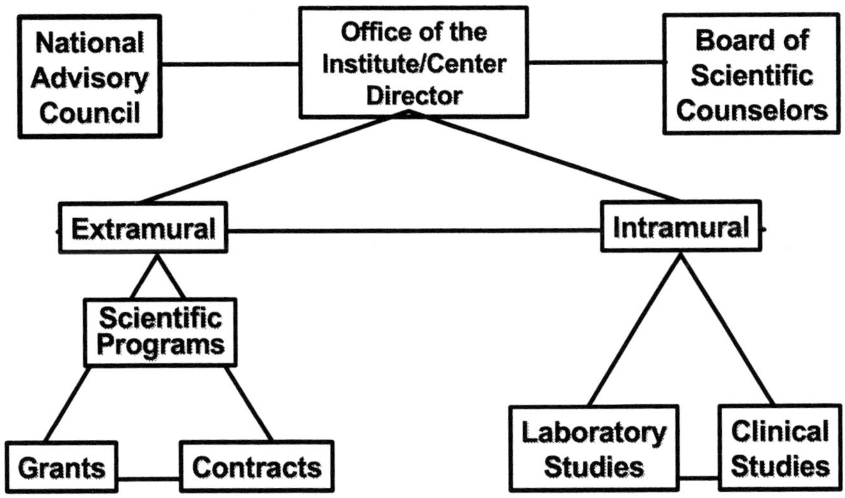

FIGURE 3-1 A typical NIH institute or center. Source: C. Thompson, NIEHS, presentation to National Research Council committee, April 25, 2002.

Each year, individual NIEHS divisions undertake a planning process to identify emphases for the coming fiscal year. The planning process takes into account progress in the scientific community, the investments of other agencies, transinstitute activities, congressional directives, and the views of program staff. Changes in emphases may be reflected in changes in the general NIEHS program announcement. The institute may also decide to release a specific RFA for a one-time competition in a well-defined scientific field.

Opportunity Communication

Program announcements and RFAs are released through standard federal publications and on the institute's Web site. This is similar to mechanisms used by NSF. Competing grant applications are received for three review cycles per year.

Proposal Review

The vast majority of the 40,000 proposals received by NIH are unsolicited and reflect the judgment of individual investigators as to their interests and the best opportunities for outstanding science. Overall, 25-30% of the proposals are funded.

Two review processes are carried out by the institutes. Unsolicited proposals and responses to the general program announcements are reviewed through the Center for Scientific Review (CSR). Proposals that respond to RFAs are reviewed within the specific institute.

The two processes use equivalent criteria for scientific merit, including the following:

- **Significance.** Does the study address an important problem? How will scientific knowledge be advanced?
- **Approach.** Are design and methods well developed and appropriate? Are problems addressed?
- **Innovation.** Are there novel concepts or approaches? Are the aims original and innovative?
- **Investigator.** Is the investigator appropriately trained?
- **Environment.** Does the scientific environment contribute to the probability of success? Are there unique features of the scientific environment?

CSR refers unsolicited proposals and responses to general program announcements to a scientific review group (SRG). SRGs are defined by specific guidelines often related to scientific disciplines or fields. The SRG provides the initial scientific review and makes recommendations for the appropriate level of support and duration of the award. It provides priority scores and percentiles for the upper half of the proposals, leaves the lower half unscored, and recommends others for deferral. CSR then refers each of the proposals to a specific institute on the basis of the institute's mission and stated interests. The institute's national advisory council (NAC) assesses the quality of the SRG review, makes recommendations to the institute staff on funding, evaluates program priorities and relevance to the institute's mission, and advises on policy. The NAC may concur with the SRGs, modify the rankings, or defer proposals for re-review. The institute director and staff then determine the specific awards to be funded on the basis of scientific merit, contribution to program balance, and availability of funds.

Proposals that respond to the RFAs from specific institutes are reviewed by institute review groups (IRGs) convened by the particular institute.

IRGs operate in a manner similar to that of the SRGs. However, the institute staff have substantial involvement in the choice of the reviewers and the amount of funds set aside for a particular number of awards. The recommendations of the IRGs are then reviewed by the institutes' scientific advisory council.

Implementation

The various NIH grant mechanisms provide for an arm's-length relationship between the investigator, his or her institution, and the institutes—similar to that in NSF. However, in contrast with NSF, the institutes use cooperative agreements and contracts for scientific services provided by user facilities or database operations. Cooperative agreements and contracts provide institute staff with considerably more influence in the conduct of research. In all cases, institute staff may choose to visit an investigator during the course of a project.

Evaluation

Individual projects are evaluated on the basis of publications in refereed journals, presentations at professional meetings, and whether students participate in the research. Projects are more highly valued if students participate. The success of an institute or division in supporting its mission or in developing a new scientific field is evaluated by the institute's national advisory council or its board of scientific counselors.

Dissemination of Results

Like NSF-supported research, NIH-supported research is disseminated through the normal mechanisms of the research community: journals and professional meetings. However, NIH planning meetings also bring together researchers and program officers to report on their work and to provide interactions that may result in future proposals. Program officers often are involved in interagency planning efforts that benefit from the newest research and development results.

DEPARTMENT OF ENERGY OFFICE OF BIOLOGICAL AND ENVIRONMENTAL RESEARCH

The Office of Biological and Environmental Research (BER) is a division of the Office of Science (SC) in the Department of Energy (DOE). BER supports basic and applied scientific research at universities and the DOE national laboratories. BER's mission arises from DOE's goal to understand and mitigate the environmental consequences of its energy and national-security missions. BER also carries out DOE's mission to develop and extend the frontiers of nuclear medicine (Elwood 2002).

BER has substantial activities that are important elements of interagency programs, such as those in long-term climate science, the human genome program, and proteomics. BER supports the operation of unique facilities at DOE national laboratories, such as high-field magnetic resonance and beamlines at x-ray synchrotron sources. Those facilities are available to university researchers supported by DOE and other agencies; access to them is determined by independent peer-review committees commissioned by the facilities and the DOE programs.

BER uses general program announcements for unsolicited proposals and for program-specific RFAs. It averages 11 such solicitations per year; these do not include joint solicitations with other agencies. The BER program budget was about $554 million in FY 2002; its university program was less than half of that amount. Hence, the peer-review process involves far fewer proposals than that of NSF or NIH, by at least a factor of 10.

Planning

BER considers a number of factors in initiating and carrying out its programs (Elwood 2002):

- Filling a national need—DOE is one of the principal federal science agencies.
- Filling a DOE mission need—DOE's national-security and energy missions are driven by science and technology.
- Filling opportunities consistent with DOE capabilities and mission.
- Filling critical scientific gaps that take advantage of DOE's physical infrastructure.
- Filling a congressional mandate.

BER uses a number of mechanisms to define program priorities: its Federal Advisory Committee Act (FACA) advisory committee, commissioned scientific assessments, scientific workshops, program-review findings, and recommendations from and staff interactions with the scientific community.

Opportunity Communication

The continuing-program announcement for the entire SC is published once a year through the *Federal Register* and the SC Web site. BER-specific solicitations are publicized in trade and professional journals, in the *Federal Register,* and on the SC Web site. Those announcements encompass both grants and cooperative agreements. They may involve a preproposal phase. University-laboratory collaboration may be encouraged; the preproposal enables the program manager to suggest useful collaborations.

Proposal Review

BER carries out peer review using a number of mechanisms: individual mail reviews, ad hoc committees, and standing committees as appropriate or required by legislation. Each application is reviewed by at least three qualified reviewers, and conflict-of-interest issues are overseen by the program manager.

BER uses the following criteria:

- Scientific or technical merit or the educational benefits of the project.
- Appropriateness of methods or approach.
- Competence of applicants and adequacy of proposed resources.
- Reasonableness and appropriateness of proposed budget.
- Other factors established in the solicitation, including program policy factors, such as program balance.

The BER program managers are responsible for determining which proposals are relevant to the agency's mission, but the program managers may ask reviewers for their recommendations. If the solicitation is expected to generate large numbers of applications, preproposals are encouraged; this step does not preclude submission of a full proposal or the normal merit-review process. Preproposals are reviewed by program managers and, if appropriate, by a panel of experts knowledgeable in the subject and aware of the program mission.

Implementation

The grant mechanism keeps the BER program official at arm's length during execution of the grant. But as in NSF, the DOE program manager visits investigators at universities and the national laboratories regularly to stay informed about progress and to support appropriate interactions among the investigators. BER programs also bring university and laboratory investigators together at regular meetings, which support implementation, evaluation, and planning efforts.

Evaluation

BER carries out retrospective reviews of its research in the grant programs and in its laboratory research. The information gained from the reviews is used to guide future program decisions as to new opportunities, program and project continuation, and balance and direction. The reviews are carried out by the FACA advisory committee, but also include the use of the JASON organization and the National Research Council's boards and committees. As in NIEHS, BER program officers participate in interagency planning efforts that benefit from early use of current R&D results.

Dissemination of Results

Results are disseminated primarily through publications and meeting presentations. BER-sponsored investigator meetings also serve as mechanisms for dissemination of results.

U.S. DEPARTMENT OF AGRICULTURE NATIONAL RESEARCH INITIATIVE COMPETITIVE GRANTS PROGRAM

The U.S. Department of Agriculture (USDA) National Research Initiative (NRI) Competitive Grants Program was established in 1991 as an element of the Cooperative State Research, Education and Extension Service

(CREES). The purpose of NRI is to support high-priority fundamental and mission-linked research of importance in the biologic, environmental, physical, and social sciences relevant to agriculture, food, and the environment. CREES reports to the USDA under secretary for research, education, and economics. The under secretary chairs the NRI Board of Directors, which oversees NRI policy (Johnson 2002).

In 2002, NRI's budget was about $120 million. Virtually all the NRI budget is devoted to grants. The grants support individual research projects at large and small colleges and universities and federal and private laboratories. NRI also funds conferences, a postdoctoral fellowship program, and a young investigator program, and it provides equipment grants. NRI is responsible for the Experimental Program to Stimulate Competitive Research (EPSCOR) program in USDA.

NRI publishes an annual program description that identifies current research opportunities in eight major topics:

- Natural resources and the environment.
- Nutrition, food safety, and health.
- Animals.
- Biology and management of pest and beneficial organisms.
- Plants.
- Markets, trade, and development.
- Enhancing value and use of agricultural and forest products.
- Agricultural systems research.

The program description also identifies strategic issues that crosscut the research topics and reflect research opportunities as they arise on the basis of the needs of the department, interagency opportunities, and developments in the scientific community. In 2002, two strategic issues were identified: agricultural security and safety through functional genomics, and new and re-emerging disease and pest threats. Although the program description provides guidance as to interests of the department and format for proposals, the proposals are considered to be unsolicited from a procurement perspective. NRI on occasion publishes directed program announcements for specific proposals.

Planning

In large measure, the annual program description summarizes the results of the annual planning efforts of the NRI chief scientist and scientific staff.

As in other agencies, scientific opportunities are identified through the proposal submissions of the research communities and the response and internal discussions of the NRI peer-review panels. Staff participation in interagency coordination committees also sets the planning horizon for the program. The staff meet regularly with commodity groups and agricultural coalition organizations that are important stakeholders of the USDA. Congress also provides input through the NRI budget process.

Opportunity Communication

The NRI program description is made available in printed form and on the Internet. It is also published in the *Federal Register* and made available at scientific meetings and conferences. Each year, NRI conducts "grantsmanship" workshops to familiarize applicants and administrators with NRI's philosophy, directives, and procedures. NRI staff have specifically presented at workshops for EPSCOR, historically black colleges and universities, and Hispanic-serving institutions.

Proposal Review

NRI receives about 3,000 proposals per year. The proposals are reviewed by panels (28 panels in FY 2000). More than 9,000 scientists contribute to the annual NRI review process. The NRI staff identify the panel members by using criteria that include relevant scientific knowledge, educational background, experience, and professional stature. Other considerations involve balancing diversity in geographic region, type of institution, rank, gender, and minority-group status. NRI staff members also ensure that conflict-of-interest policies are enforced.

At a panel meeting, each project is reported on by three panelists. The primary reviewer provides an overview and summary of the proposal's strengths and weaknesses. A secondary reviewer provides additional comments, and a reader summarizes a set of four to six ad hoc reviews that are provided by other members of the panel or from scientists outside the panel. Each project is ranked in one of six categories, the lowest of which is "do not fund." For each project, the panel provides a summary of positive and negative aspects and a synthesis of the discussion in the larger context of panel considerations. On the final day of the panel meeting, the projects are re-ranked by revisiting the categories and providing a numerical rank order for the top 25%. Awards are based on the panel's ranking. The NRI staff finalize the budgets but do not adjust individual project budgets. Each project investigator receives the reviews, the panel summary, and the relative ranking of the project.

Implementation

As in other grant programs, initial terms and conditions govern the performance of the grant. There is no official mechanism for NRI program managers to direct principal investigators during the term of the grant.

Evaluation

The National Research Council has provided reviews of NRI in 1994 and 2000 (NRC 1994; NRC 2000). NRI also evaluates itself against CREES's Government Performance and Results Act (GPRA) outcomes by providing appropriate descriptions of project impacts on those outcomes. The NRI Board of Directors provides oversight of NRI policy retrospectively and prospectively.

Dissemination of Results

As in other programs, publication in refereed journals, presentations at scientific conferences, and graduate students are major mechanisms for dissemination of results. The NRI staff present results at meetings with various USDA stakeholders that convey the impact of NRI research. NRI also uses the extension resources of CREES to prepare and distribute one-page summaries of particular NRI results. The NRI Board of Directors—which is composed of the under secretary and four administrators in the Research, Education, and Economics directorate—also serves as a forum for information and exchange of NRI results.

NATIONAL AERONAUTICS AND SPACE ADMINISTRATION

The National Aeronautics and Space Administration (NASA) is one of the largest federal science and technology agencies. It accomplishes most of its mission through its major laboratories and contracts with private industry. However, it provides substantial support for university researchers through contracts, cooperative agreements, and grants, which are awarded through a merit-based peer-review process that includes the participation of the scientific community (B. Bennett, NASA, personal commun., June 2002).

Three NASA divisions support peer-reviewed research: Space Science, Earth Science, and Biological and Physical Science Research. The FY 2002 budget for those three divisions was about $5.81 billion. The divisions use similar peer-review processes that conform with overall NASA guidance. Each uses the NASA research announcement; an announcement may identify a division's broad topics of interest as guidance for unsolicited proposals or may solicit proposals in specific research topics. The NASA Advisory Council (NAC), a FACA committee, has established standing committees for each of the divisions with a strong focus on scientific direction. The chairs of the National Research Council's Space Studies Board (SSB) and Aeronautics and Space Engineering Board (ASEB) are ex officio members of the Advisory Council (B. Bennett, NASA, personal commun., June 2002).

Planning

Each of NASA's research divisions contributes to the NASA strategic plan. With respect to the identification and priorities of research, the development of NASA research announcements is an opportunity for interaction with and feedback from the scientific community. The standing committees of the NAC also undertake program reviews and planning activities that affect the divisions' plans. Congressional direction through the budget process is incorporated into plans. The studies of the Research Council's SSB and the ASEB are often used in research planning (NASA 2001).

Opportunity Communication

NASA research announcements are broadly communicated on the NASA Web site and in print through the *Commerce Business Daily* and the *Federal Register*. The involvement of the community in the development of the announcement is also part of NASA's communication effort (NASA 2001).

Proposal Review

NASA uses both letter and panel reviews, depending on the decision of the program manager. NASA has contracted with a support-service contractor to handle the logistics of the peer-review process. All divisions that use peer review work with the same contractor to provide consistency of interaction with proposers and reviewers. The contractor maintains proposal and reviewer databases. The program manager suggests reviewers and works with the contractor to ensure conformance with conflict-of-interest rules.

The principal elements considered in evaluation are intrinsic merit, relevance to NASA's objectives, and cost. The evaluation of intrinsic merit takes into account overall scientific or technical merit, qualifications of the investigator and other key personnel, institutional resources, experience critical to objectives, plans for education and outreach, and overall standing among other proposals. The intrinsic-merit review does not generally depend on cost or programmatic relevance, but reviewers may be asked for comments on cost and relevance.

Proposals for science that depend on space transport will be evaluated for engineering and management by a panel that includes government and contractor reviewers. This review judges proposals on the feasibility and complexity of accomplishing project goals; it also provides an estimate of the total cost of flight-hardware development.

The NASA program manager has considerable latitude in determining how a project will be evaluated and the funding that will be made available for a project. For complex projects, the program manager may not choose the procurement mechanism (grant, cooperative agreement, or contract) until the review process is completed (NASA 1999).

Implementation

A NASA program manager's relationship with grantees is subject to the same constraints as in other science agencies. However, the connection between grant-supported research and the missions of NASA divisions results in considerable interaction through formal and informal meetings, including professional-society meetings. The program manager often takes an active role in organizing those meetings (NASA 1999).

Evaluation

The full NAC reviews the research divisions annually to assess their performance against their GPRA measures. The NAC standing committees do retrospective reviews of specific elements of the divisions. NASA also establishes external ad hoc review committees. For example, the Biological and Physical Research Division has established the Research Maximization and Prioritization Task Force to review and assist in planning for research that will use the International Space Station. The National Research Council has also reviewed specific NASA programs; the SSB and the ASEB have standing committees that often review program activities (NASA 1999).

Dissemination of Results

Of all the federal science and technology agencies, NASA is the only one that has a statutory mandate and receives specific funding for publicizing its mission and its results to both the technical community and the general public. Thus, although NASA-supported science is disseminated through journals and professional meetings, it may also be chosen for broader distribution to the public media. That not only benefits the research and NASA but also enhances the likelihood of a broader research impact (NASA 1999).

NATIONAL OCEANIC AND ATMOSPHERIC ADMINISTRATION

The mission of the National Oceanic and Atmospheric Administration (NOAA) is to describe and predict changes in the earth's environment and to conserve and manage wisely the nation's coastal and marine resources to ensure sustainable economic opportunities (NOAA, unpublished material, 1998). The largest divisions of NOAA are the National Weather Service (NWS), the National Marine Fisheries Service (NMFS), the Coastal Ocean Service (COS), and the Office of Ocean and Atmospheric Research (OAR).

The agency accomplishes its mission through inhouse and extramural activities. Although some extramural research funds are managed by NMFS, COS, and NWS, most of the inhouse and extramural research efforts are housed in the OAR. Inhouse research—programmatically divided into oceans, weather, climate, and atmosphere at OAR—is carried out at 12 laboratories. The extramural programs include the National Sea Grant College Program,[1] the National Undersea Research Program, the Joint Institute Program, the Arctic Research Program, and the Office of Global Change Program. OAR's FY 2002 budget was about $340 million.

Planning

Each of NOAA's units contributes to the agency's strategic plan. OAR receives considerable input from the nation's major research universities, environmental managers, and the general public. Much of this input is coordinated by the National Sea Grant College Program network in the coastal states, by the National Underwater Research Program, and by the joint institutes. Input is coordinated by three OAR Assistant Deputies for agency cross-cutting and interdisciplinary research activities.

Opportunity Communications

All research opportunities are broadly communicated in the *Federal Register*, the *Commerce Business Daily*, and direct mailings and on NOAA, joint institutes, and sea-grant Web sites.

Proposal Review

NOAA uses letter reviews and panel reviews for the various funding programs. The 31 Sea Grant College programs receive and process more

---

[1]The National Sea Grant College Program, created in 1966, established a partnership between NOAA and universities to encourage the development of sea-grant institutions for the purpose of engaging in research, education, outreach, and technology transfer in an effort to encourage stewardship of the nation's marine resources.

proposals than any other NOAA division. All programs have uniform guidelines and oversight for peer reviewers, although individual programs maintain their own databases.

Reviewers are asked to evaluate proposals on the basis of scientific merit and whether the proposed work addresses the mission of NOAA and in particular the mission of the division. Cost, opportunities for outreach activities, qualifications of investigators and key scientific personnel, and institutional research infrastructure support are also considered.

Implementation

The extramural grant programs ensure an arm's-length relationship between researchers and program officers. However, as in other mission agencies, there are strong incentives for visits and exchange to stay abreast of program issues and to ensure prompt input into the research results. The regional character of the National Sea Grant College Program also provides opportunities for regular interaction between researchers at different institutions.

Evaluation

The Joint Institute Program, the Arctic Research Program, and the Office of Global Change Program are all evaluated through regular peer-review visits. Detailed evaluation criteria and performance benchmarks have been developed for the National Sea Grant College Program. The criteria include the following components: effective program planning; organizing and managing for success with meritorious project selections, recruiting of the best talent available, and meritorious institutional program components; and producing significant results. Program management must ensure that consistent production of significant results will have widespread economic or social benefit, contribute to science and engineering, and address the high-priority needs of the state and the nation.

Dissemination of Results

As in other agencies, NOAA research is disseminated through workshops, scientific meetings, and journal articles. Because the National Sea Grant College Program's mission includes research, outreach, and educational components, it has developed criteria and benchmarks for "connecting with the user."

CONCLUSIONS

- The agencies described above have research-management processes to plan, solicit, select, and evaluate extramural research activities. All use a competitive peer-review process that draws expert reviewers from the scientific community, and they have processes for screening the reviewers for expertise and conflicts of interest.
- All agencies can use a number of procurement mechanisms, although grants are the principal mechanism for researchers based in universities. Grants require a hands-off relationship between researchers and agency staff; however, all the agencies use some form of meeting or conference to assess progress during the grant period. The meetings can also have other functions: they permit interaction among scientists and between scientists and a broad array of agency staff, resulting in prompt research impact related to agency missions. In addition, the meetings support research collaboration and planning efforts for future research directions and solicitations.
- The agencies have all been required to address GPRA requirements for strategic planning and the development of metrics. They recognize that it is difficult to identify quantitative outcome metrics on problem-driven research even when mission-related, disciplinary topics can be identified.
- In establishing and managing extramural research, the agencies deal with the same issues, and many face the same kind of mission-imperatives that underlie the STAR program. However, most of the agencies described here administer much larger budgets than the STAR program. For example, the NIEHS budget is about 15 times the STAR budget; NSF's environmental programs are nearly 9 times larger.
- EPA appears to have benefited from the STAR program's collaborative efforts with other agencies. First, the joint programs have resulted in more EPA-relevant research than STAR would have been able to fund alone, with the joint research mutually benefiting both EPA and the partnering agency. Second, STAR has been able to accelerate the development of its management processes by learning from its partners. For example, like NIEHS, it has established a separate organization to conduct peer review. At the same time, the well-developed EPA Office of Research and Development planning process and STAR's strong involvement with it provide the STAR program with greater integration and relevance to the EPA mission,

at least at the process level, than is apparent in some of the other agencies described here.

- In general, the committee finds that STAR processes compare favorably with those of its peer agencies, particularly given the relative youth of STAR. In addition, the STAR program's relevancy review process is more rigorous than that of other agencies.

REFERENCES

Elwood, J. 2002. Department of Energy's Biological and Environmental Research Grants Program. Presentation at the Second Meeting on the Review of EPA's Research Grants Program, April 25, 2002, Washington, DC.

ERE (Environmental Research and Education). 2003. What is ERE. Environmental Research and Education, The National Science Foundation, Arlington, VA [Online]. Available: http://www.nsf.gov/geo/ere/ereweb/ what.cfm [accessed Jan. 10, 2003].

Firth, P. 2002. National Science Foundation Research Grants Program. Presentation at the Second Meeting on the Review of EPA's Research Grants Program, April 25, 2002, Washington, DC.

Johnson, P. 2002. United States Department of Agriculture, National Research Initiative. Presentation at the Second Meeting on the Review of EPA's Research Grants Program, April 25, 2002, Washington, DC.

NASA (National Aeronautics and Space Administration). 1999. Office Work Instruction. Research Solicitation, Evaluation, and Selection. National Aeronautics and Space Administration. [Online]. Available: http://hqiso9000.hq.nasa.gov [accessed July 2, 2002].

NASA (National Aeronautics and Space Administration). 2001. Office Work Instruction. NASA Research Announcement (NRA) for R&D Investigations. National Aeronautics and Space Administration. [Online]. Available: http://hqiso9000.hq.nasa.gov [accessed July 2, 2002].

NRC (National Research Council). 1994. Investing in the National Research Initiative: An Update of the Competitive Grants Program in the U.S. Department of Agriculture. Washington, DC: National Academy Press.

NRC (National Research Council). 2000. National Research Initiative: A Vital Competitive Grants Program in Food, Fiber, and Natural-Resources Research. Washington, DC: National Academy Press.

Thompson, C. 2002. NIEHS Research Grants Program. Presentation at the Second Meeting on the Review of EPA's Research Grants Program, April 25, 2002, Washington, DC.

4

Measure for Measure

The committee was charged with assessing the Environmental Protection Agency (EPA) Science To Achieve Results (STAR) program's scientific merit, its demonstrated or potential impact on the agency's policies or decisions, and other program benefits that are relevant to the agency's mission. Assessment implies measurement. EPA, the Office of Management and Budget, and Congress are intensely focused on using metrics as a means of gauging the value of federal research programs. As a result, officials in EPA's Office of Research and Development (ORD) urged the committee to develop and use metrics in its evaluation of the STAR program.

Because of the breadth and many dimensions of the committee's task, the committee considered a broad range of metrics in its evaluation of the STAR program. Clearly a one-size metric will not fit all aspects of the program. This chapter provides a foundation for metrics, addressing what they are, how they are used in evaluation, and some considerations that should be given to their selection and use. The motivation for the emphasis on metrics, stemming from the Government Performance and Results Act (GPRA), is discussed. Finally, bibliometric analysis, a common form of metric, is discussed in relation to the STAR program. Appendix C contains examples of metrics used by federal agencies, academe, and state governments in evaluating their research programs.

WHAT ARE METRICS?

Geisler (2000) defines metrics for the evaluation of a research program as "a system of measurement that includes, (1) the item or object that is being measured; (2) units to be measured, also referred to as 'standard units'; and (3) value of a unit as compared to other units of reference." Geisler (2000) goes on to clarify this definition as follows:

> A refinement of the definition of metric extends it to include: (a) the item measured (*what* we are measuring), (b) units of measurement (*how* we measure), and (c) the inherent value associated with the metric (*why* we measure, or what we intend to achieve by this measurement). So, for instance, the metric *peer review* includes the item measured (scientific outcomes), the unit of measurement (subjective assessment), and inherent value (performance and productivity of scientists, engineers, and S&T units).

Types of Metrics

Metrics may be classified as quantitative, semiquantitative, and qualitative. For the purpose of this report, the committee characterizes metrics as quantitative or qualitative, grouping the semiquantitative measures with the qualitative.

Quantitative measures, such as the number of peer-reviewed publications resulting from a grant, have the desirable attributes of public availability and reproducibility. A drawback to quantitative metrics is that they tend to be reductive or one-dimensional, measuring a single quantity. As a result, quantitative metrics, although outwardly simpler to use, are not necessarily more informative than qualitative metrics. Quantitative metrics tend to be more useful at lower levels of evaluation, when information tends to be more discrete, such as a review of a specific grant or center, but become less useful as one evaluates higher levels of integration, such as a review of the entire STAR program.

Qualitative metrics have the advantage of being multidimensional, that is, of comprising an intricate and complex set of measures. Therefore, qualitative metrics are more useful for evaluating higher levels of integration, such as an entire research program. Qualitative metrics can have numerical components; for instance, in reviewing grants, a scoring system of 1 to 5 is commonly used, in which the numbers represent such labels as "excellent" and "very good." They do not have many of the characteristics of normal quantitative evaluations; example, a 4 is clearly better than a 2, but it is not necessarily twice as good.

Process and Product Metrics

Metrics may also be classified on the basis of various components of a research program that they are used to evaluate. For instance, Cozzens (2002) defines four types of metrics: input, or money allocated and spent; throughput, or project activities; output, or publications, people, and products; and outcome, or user satisfaction.

The committee chose to use the terms *process metrics* and *product metrics* for evaluating the STAR program. Process metrics are used to evaluate the operation or procedures of the STAR program, such as peer review. Product metrics describe the outputs from the program, such as the number of reports, or the influence or effect of the program. In a crude way, process metrics represent internal program assessments and product metrics represent external program assessments.

The emphasis on product metrics is understandable, inasmuch as it is important that programs focus on what they are accomplishing, but process metrics are also important, particularly in warning of possible problems in a research program. A substantial period of time may pass before research programs generate enough products to support a comprehensive review. If such a review indicates serious problems in research results, the processes that led to the problems probably occurred many years earlier. Research managers consider good processes (such as a good peer-review process) to be necessary but not sufficient for ensuring good products.

Attributes of Metrics

It is important that metrics developed for evaluating a program fit together to provide a clear, accurate, comprehensive, and coherent picture. The committee considers some important attributes of metrics to be the following:

- **Meaningful.** The metrics should be related to topics that the intended audience cares about. The first step in any evaluation should be to identify the audience for which it is being conducted.

- **Simple.** The metrics and their measurements should be expressed in as simple and concise terms as possible so that the audience clearly understands what is being measured and what the results of the measurement are.
- **Integrated.** The metrics should fit together to provide a coherent picture of the program being evaluated. The focus should be on performance goals and baseline statements in a way that provides a comprehensive recognition of accomplishments and the identification of information gaps.
- **Aligned.** The metrics should be solidly aligned and accurately reflect the overall program and agency goals.
- **Outcome-oriented.** Although some metrics focus on process, the best focus on desired environmental benefits, not only on the completion of tasks.
- **Consistent.** If multiple programs are being evaluated or the same program is being evaluated over time, the same metrics should be used for the different programs and the different times.
- **Cost-effective.** Program evaluations can be expensive in the resources required to support them and in the disruption that they can cause to the program being evaluated. The benefits of the information that the metrics provide should be commensurate with the costs required to collect it. It is most cost-effective to use information that is being (or should be) collected to support continuing effective program management.
- **Appropriately timed and timely.** Some metrics require information to be collected on a continuing basis, others require information to be collected annually or even less frequently. For an evaluation to be accurate, the information has to be up to date. In some cases, that can influence when the evaluation is undertaken to ensure that it is based on current information.
- **Accurate.** The metrics should promote the collection of information that accurately measures how the program is doing. Inaccurate metrics can seriously skew a program's performance, inducing it to emphasize accomplishments that show up well according to the metric but that are irrelevant to and perhaps even inconsistent with the program's goals.

USE OF METRICS IN EVALUATION

The selection of the appropriate metrics necessary to evaluate a research project or program is constrained by the nature of the evaluation, including the purpose of the evaluation (for example, the intended audience and why the evaluation is being prepared), the type of research activity being evaluated, and the nature of the results that are of interest.

A research program like STAR is multidimensional and therefore should be evaluated on several levels, each level requiring its own set of metrics. The committee considers that a reasonable and logical approach to addressing such complexity in the STAR program encompasses four levels of review (see Figure 4-1). Level 1 focuses on the individual STAR projects and the higher levels of review are more integrative, focusing on research topics (level 2), the overall STAR program (level 3), and the EPA research program (level 4). For each level, two potential types of review may occur: process review and product review. Each type of review draws on metrics pertaining to the level of review. At each level of review, independent experts with appropriate scientific, management, and policy backgrounds would conduct the reviews.

Level 1. The first level of review pertains to an individual research project. Typically, the project officer is responsible for carrying out the process review at this level, which usually includes such issues as ensuring that the work is being conducted on schedule by the appropriate investigators, that annual and other reports are being submitted as required, and that other federal administrative requirements are being satisfied.

A product review is also sometimes carried out for individual research projects. Such a product, or substantive, review is the norm for research being conducted under a contract or cooperative agreement. It is less common for research conducted under a grant. If the project officer has sufficient technical expertise, he or she may conduct both the product review and the process review. However, some organizations bring in an outside expert or even, if the project is unusually large and complex, a team of experts representing the diverse disciplines the project is supposed to be incorporating. At level 1, it is reasonable to conduct both process and product reviews.

At this level, process reviews should be undertaken by people who are familiar with the administrative requirements and the commitments made in the research agreement. Product reviews should be conducted by people who are experienced in conducting research pertaining to the topic and have a knowledge of good research methods.

Level 2. The second level of review pertains to a topic, or a group of projects addressing the same general subject (such as particulate matter or

Level of Review	Focus of Review	Type of Review: Process	Type of Review: Product	Appropriate Mechanism For Review
1	Individual Project			Experienced individual such as a STAR project officer
2	Topic Area			An independent technical committee comprised of experts in the topic area who do not have a conflict of interest
3	STAR Program			An independent technical committee comprised of persons with experience in government operations, research Management, and environmental issues
4	EPA Research Program			An independent technical committee comprised of persons, from within and outside of government, who have experience in government operations and research management

FIGURE 4-1 Levels of review.

ecologic indicators). This is often the most efficient level for conducting product reviews focusing on the substance of the research. Such reviews are typically conducted by an external review committee comprising experts in the subject being reviewed. The experts should have a good overview of the topic and of the scientific research being conducted in the topic. The experts are in a position to judge the quality of the research being conducted, the success that EPA has had in identifying high-priority research that will fill important gaps in the scientific community's understanding of the topic, and whether the research being sponsored duplicates work that is being or has been done elsewhere.

Such reviews are most productively undertaken when there is a sufficient body of research results to provide a basis for drawing informed conclusions on the issues being addressed. Because of their cost and the disruption that they can cause, they should be undertaken infrequently. These reviews often also consider process issues, but process issues are more efficiently addressed at levels 3 and 4.

Level 3. The third level of review focuses on the operation and management of the research program and its products. It addresses such issues as how effectively the program communicates research opportunities, how objectively it evaluates proposals, how efficiently it awards grants, and how carefully it monitors projects.

The third level of review is often conducted by a panel of experts, including people who have experience in managing research programs, who are familiar with other research programs that complement or compete with the one being reviewed, who are familiar with the institution where the research program is conducted, and, in the case of government research programs, who are familiar with government operating and administrative procedures.

Level 4. The fourth level of review is often undertaken by an expert committee that comprises people with the same characteristics and experience as those in the level 3 review. This committee includes individuals from both within and outside of government. The focus of the fourth level of review, however, is broader. It is concerned less with how the program operates internally than how it is related to the broader institution and how effective it is in responding to the information needs of the organization and other potential users of the research results. The fourth level addresses such issues as whether the research organization has properly identified its clients, how well the research planning process and the definition of specific research topics include the perspectives of potential users, how effective the research organization is in keeping potential users "plugged in" to the research as it progresses, and how well it disseminates research results to potential users. If the institution sponsors multiple research efforts, as is the case with EPA and its various research centers and laboratories, the fourth level of review also addresses how well and efficiently the efforts are coordinated.

There can be other levels of review. For instance, expert committees are sometimes established to review all the information pertaining to a topic of particular interest. Such committees do not focus on the work conducted or sponsored by one institution but rather review information that has been collected in government, in the academic and nonprofit communities, by private businesses, and in foreign countries. Often, the charge given to an expert committee overlaps two or more levels of review. However, asking a committee established at one level to conduct a review that is more appropriate for another level is likely to result in inefficiencies and may not provide the degree of insight desired.

The present committee's own evaluation focused on the overall STAR program—that is, level 3—considering the operation and management of the STAR research program. But the committee also considered the STAR program in relation to ORD and to EPA as a whole (level 4). (The committee's evaluation of the STAR program is addressed in Chapter 5.)

Caveats Regarding the Use of Metrics

The committee urges caution with respect to the use of metrics in evaluating research programs, because there is an inherent danger of measuring too much and too often. As Geisler (2000) states, "metrics selected should be able to measure what the evaluators wish to be measured. They should be intimately linked to the objectives and motives of the evaluators ... If this criterion is not met, the metrics become a set of irrelevant and useless quantities." Many others have issued cautions regarding the use and application of metrics. For instance, the National Research Council Committee on Science, Engineering, and Public Policy (COSEPUP) sounded the following warning (NRC 1999):

> It is important to choose measures well and use them efficiently to minimize non-productive efforts. The metrics used also will change the behavior of the people being measured. For example, in basic research if you measure relatively unimportant indicators, such as the number of publications per researcher instead of the quality of those publications, you will foster activities that may not be very productive or useful to the organization. A successful performance assessment program will both encourage positive behavior and discourage negative behavior.

BIBLIOMETRIC ANALYSIS

One evaluation tool that has gained much currency is bibliometric analysis. The committee commissioned such an analysis for a small subset of STAR-funded research to help to assess the utility of this technique in evaluating the STAR program (IISCO, Atlanta, GA, unpublished material, 2002).

Bibliometric analysis is based on the premise that a researcher's work has value when it is judged by peers to have merit (NRC 1999), and it is commonly used in research evaluations because it seeks to measure research productivity by quantifying publication outputs and citations. The popularity of bibliometrics stems from its quantitative nature; it lends itself readily to "ranking." It can be used in the identification of productive people, institutions, and even countries and in the charting of trends in research (for example, *endocrine disruptors* is a term that has recently come into fashion). Bibliometric analysis is relatively inexpensive (Geisler 2000).

Bibliometrics has a number of limitations. When a given author's publications are counted, the extent to which specific publications are derived from a particular research project is rarely clear. Furthermore, some disciplines publish more than others, and some researchers publish more than others.

It is also important to consider the quality of the journal within each discipline in which an article is published. Articles accepted by a prestigious journal can have more influence and should be more highly valued in a bibliometric analysis than articles published in less prestigious journals.

Some materials are covered less fully than others in bibliometrics. Published articles are only one measure of the output of scientific activity, but bibliometrics does not cover, for example, electronic communications, chapters in books, and abstracts.

Additional limitations include the current inability to screen text. Word references are typically based on key words; citations of multiauthor articles tend to be truncated to two or three authors; and in highly collaborative, cross-disciplinary applied research (such as that sponsored by the STAR program), results are published in diverse journals—for instance, research addressing environmental causes of childhood asthma can appear in journals dealing with buildings, general medicine, toxicology, epidemiology, molecular biology, agriculture, or social science (Geisler 2000). Thus the particular abstracts included in the analysis and the key words used to search the abstracts can have an important influence on the publication count. In the analysis undertaken for the committee (discussed below), some of the articles published by some of the researchers did not appear; the journals in which the articles were published were not included in the collections of abstracts that were searched, or the abstracts did not include the specific key words that were used for the search.

Cross-disciplinary comparisons of counts of publications or citations can also be misleading because researchers in some disciplines tend to publish journal articles more frequently than those in others. If a discipline supports a number of journals, essentially the same article can be repeated by someone who knows how to play the publication game.

The citation-analysis component of bibliometric analysis can avoid some of the problems just noted; for instance, an article published in a prestigious journal is likely to be cited more frequently than one published in an obscure journal. However, citation analyses have their own limitations. One is that they can be biased by an inordinate amount of self-citation and citations by "friends" (Geisler 2000). A high rate of citation does not necessarily provide a measure of quality. For example, an article that contains a serious error or is otherwise controversial may be cited frequently by researchers eager to demonstrate its failings.

Cross-disciplinary comparisons can pose a problem in citation analyses. Disciplines have their own citation traditions. In some fields, such as internal medicine, the senior collaborator is the last author listed and therefore often does not appear in citations that list only the first three authors. In other fields, the primary writer is the first author and the senior collaborator the second author. In mathematics, it is traditional to list authors alphabetically. Interpreting a bibliometric or citation analysis properly requires knowledge of the traditions of a particular research field.

Finally, an article on a topic that many researchers are working on may be widely cited, whereas an article on a new topic or filling an information gap that is being ignored by other researchers may not be, even though the latter may constitute a more important contribution to the state of the science.

Geisler (2000) states that the "consensus among the critics is that the metric has some merit, but its value as a 'stand-alone' metric is doubtful." The committee agrees with that assessment and recommends the use of bibliometric analysis only to support expert reviews; review by a group knowledgeable about a specific research topic will assist in placing the results of bibliometric analysis in the context of the current state of scientific knowledge.

A bibliometric analysis commissioned by the committee analyzed results of grants awarded in response to two requests for applications: 10 ecologic-indicators grants and eight endocrine-disruptor grants funded by STAR in 1996. The assessment indicated that the number of publications by STAR grantees was comparable with that by other researchers in the fields and that the grantees were producing high-quality work. A citation analysis (see Table 4-1) indicated that the rate of citations of STAR-funded research was similar to that of other research in the field.

Overall, the committee concluded that it is essential for bibliometric analysis to be done in conjunction with expert review to assess its quality

TABLE 4-1 Mean Citations for STAR-Funded Research and Other Research in the Fields of Ecologic Indicators and Endocrine Disruptors[a]

	Ecologic Indicators		Endocrine Disruptors	
Year[b]	Other Research	STAR Research	Other Research	STAR Research
1997	10.3	10.5	13.2	37.8
1998	8.6	7	5.9	2.6
1999	5.3	7.7	7.8	6.9
2000	2.7	2.8	3.3	2.9
2001	1.1	0.7	1.2	0.4

[a]Identification of STAR-funded research publications is based on investigators' judgment.
[b]Papers published earlier can accrue more citations than those published more recently.
Source: IISCO, Atlanta, GA, unpublished material, 2002.

and relevance, inasmuch as the method does have inherent limitations. That conclusion supports a similar recommendation made by COSEPUP (NRC 1999). The committee also believes that the STAR program is too young for bibliometric analyses to be an effective metric for research funded beyond the initial years of the program.

GOVERNMENT PERFORMANCE AND RESULTS ACT

Much of the recent focus on the use of metrics to evaluate research programs stems from the Government Performance and Results Act (GPRA) of 1993. GPRA is intended to focus agency and oversight attention on the outcomes of government activities, so as to ensure the accountability of federal agencies. To that end, it requires each agency to produce three documents: a strategic plan that establishes long-term goals and objectives for a 5-year period, an annual performance plan that translates the goals of the strategic plan into annual targets, and an annual performance report that demonstrates whether an agency's targets have been met. Federal research agencies have developed various planning processes in response to GPRA.

Although GPRA offers agencies the opportunity to communicate to policy-makers and the public the rationale for and results of their research programs, it has created substantial challenges for many research agencies (GAO 1997; NRC 1999). Results of research activities are unpredictable and long-term; this places limitations on the use of roadmaps or milestones of progress. Annual performance measures tend to focus on less important findings rather than major scientific or technologic discoveries or advances (Cozzens 2000). Research agencies face other challenges with respect to the political environment surrounding GPRA. The Office of Management and Budget (OMB)wants GPRA to provide it with measures of good management at agencies. However, for research agencies, focusing on good management, although important, does not necessarily produce research results for the public (Cozzens 2000).

In 1999, Congress asked COSEPUP to provide guidance on how to evaluate federal research programs relative to GPRA. The request was in response to the difficulties that federal research agencies were having in linking results with annual investments in research.

COSEPUP concluded that federal research programs can be usefully evaluated regularly in accordance with the spirit and intent of GPRA. However, useful outcomes of basic research cannot be measured directly on an annual basis. The COSEPUP report cautioned that evaluation methods must be chosen to match the character of research and its objectives (NRC 1999). The report concludes that quality, relevance, and leadership are useful measures of the outcome of research (particularly basic research). Federal agencies should use expert review to assess the quality of the research they support, its relevance, to their missions, and its leadership (NRC 1999).

In February 2002, OMB proposed preliminary investment criteria that could be used for evaluating federal basic-research programs. The criteria—quality, relevance, and leadership—were a combination of criteria suggested by COSEPUP and by the Army Research Laboratory (ARL); ARL had selected quality, relevance, and productivity as relevant metrics for evaluating programs of basic and applied research (OMB 2002).

The importance of the OMB criteria was emphasized in a May 2002 memorandum from John H. Marburger, director of the Office of Science and Technology Policy, and Mitchell Daniels, director of OMB. The memorandum included a slightly revised set of evaluation criteria and directed agencies to use the new R&D investment criteria—covering quality, relevance, and performance—in their FY 2004 R&D budget requests (see Table 4-2).

The criteria are intended to apply to all types of R&D, including applied and basic research. However, the memo notes that the administration is aware that predicting and assessing the outcomes of basic research is never

TABLE 4-2 Office of Management and Budget Research and Development Investment Criteria

Quality.
R&D programs must justify *how* funds will be allocated to ensure quality R&D. Programs allocating funds through means other than a competitive, merit-based process must justify the exceptions and document how quality is maintained.
Relevance
R&D programs must be able to articulate *why* this investment is important, relevant, and appropriate. Programs must have well-conceived plans that identify program goals and priorities and identify linkages to national and "customer" needs.
Performance
R&D programs must have plans and management processes in place to monitor and document *how well* this investment is performing. Program managers must define appropriate outcome measures and milestones that can be used to track progress toward goals and assess whether funding should be enhanced and redirected.

easy. The extent to which the criteria are to take the place of GPRA is not clear. As the memo states, "these criteria are intended to be consistent with budget-performance integration efforts. OMB encourages agencies to make the processes they use to satisfy GPRA consistent with the goals and metrics they use to satisfy these R&D criteria. Satisfying the R&D performance criteria for a given program will serve to set and evaluate R&D performance goals for the purposes of GPRA" (OSTP/OMB 2002).

Table 4-3 compares the OMB criteria of quality, relevance, and performance with the recommendations of the COSEPUP report, EPA's Office of Research and Development strategic goals (Chapter 2), and the STAR program goals (Chapter 2). From Table 4-3, it is evident that OMB's R&D criteria are not separate from those of COSEPUP, ORD, or STAR, but rather comprise many of these other criteria or goals. Examining the OMB criteria in this context underscores the fact that the criteria encompass the objectives of EPA's mission and fall within the research criteria and goals established by COSEPUP and STAR.

To understand how metrics can be used to evaluate programs effectively, the committee reviewed metrics used by other organizations in federal and state governments and academe. Chapter 3 identified some of the

TABLE 4-3 Comparison of Research Criteria and Goals

OMB[a]	COSEPUP[b]	2001 ORD Strategic Plan[c]	1995 STAR Goals[d]
Quality	Quality		Achieve excellence in research
	World leadership	Science leadership	
Relevance	Relevance	Support agency's mission	Focus on highest-priority mission-related needs
	Progress toward practical outcomes	Integrate science and technology and to solve problems	Integrate and communicate results
		Anticipate future issues	
Performance		High performance	High accountability and integrity
			Partnerships and leveraging
	Develop and maintain human resources		Develop next generation of environmental scientists

[a]OMB 2002.
[b]NRC 1999.
[c]EPA 2001.
[d]P. Preuss, presentation to the National Research Council committee, March 18, 2002.

efforts of federal agencies, and additional efforts by other federal agencies and state governments and academe are summarized in Appendix C.

Review of the metrics discussed in Appendix C indicated that federal research programs tend to focus more on the collection of product metrics than on process metrics. In federal research programs, there tends to be a presumption that peer review is necessary to ensure a successful program. However, there also tends to be relatively little discussion as to who is responsible for conducting peer review. The Air Force, however, uses a highly quantitative evaluation process, with the reviews being conducted by the Air Force Scientific Advisory Board.

Evaluations at the state level are driven principally by economic considerations. There tends to be little targeting of specific research topics except in broad terms, such as nanotechnology. Many of the evaluations are based on surveys of participating institutions and on data routinely collected at the state level, such as numbers of students enrolled in institutions of higher learning. The Experimental Program to Stimulate Competitive Research (EPSCOR) of the National Science Foundation produces a level of standardization that allows comparison of R&D efforts across states and across time. The standardization provides consistency, an important attribute of metrics.

In the committee's view, none of the evaluation programs has identified the "silver bullet" that will provide an unambiguous measure of the quality of a research program, and many of the organizations continue to wrestle with the problem of how to evaluate their research programs effectively without imposing undue costs or disruptions.

In Chapter 5, the committee uses OMB's R&D criteria to evaluate the quality, relevance, and performance of the STAR program, as these criteria are to be used by government agencies in assessing their research programs for the FY 2004 budget (OSTP/OMB 2002). In its evaluation, the committee reviewed a large number of potential metrics used by EPA and other organizations and selected a small set to evaluate the STAR program. The metrics are classified as process or product metrics. Because the committee conducted a process-oriented review (level 3), the metrics used in the evaluation tended to be more qualitative than quantitative.

CONCLUSIONS

- On the basis of its review of numerous metrics being used to gauge research programs in and outside EPA, the committee concludes that there are no "silver bullets" when it comes to metrics. The committee concludes that expert review undertaken by a group of persons with appropriate expertise is the best method of evaluating the STAR research program. The types of experts needed will depend on the level of review being conducted. The use of expert review is supported by recommendations made by COSEPUP and the practices of federal research agencies.

- Expert review panels should be used for evaluating the processes and products of the STAR program. Both process and product reviews are important but should be conducted at the appropriate program levels. A good process is generally necessary but not sufficient to ensure a good product. Thus, product reviews are necessary to ensure that a program is producing high-quality results.
- The committee recommends that the STAR program consider establishing a structured schedule of expert reviews that has four levels: level 1, individual research projects; level 2, topics or groups of research projects; level 3, the entire STAR program; and level 4, the question of how the STAR program is related to the broader institutions of ORD and EPA. Each level should have its own set of metrics. As reviews move to higher levels of program evaluation (from level 1 to level 4), integration becomes more important, and metrics become more qualitative than quantitative.
- The expert reviews should use qualitative and quantitative metrics to support their evaluations. Both types of metrics are valuable in assessing research projects and programs. Quantitative metrics, although outwardly simpler to use, are not necessarily more informative than qualitative metrics. In fact, a numerical veneer can often be more difficult to interpret and less transparent in that it may hide unsuspected idiosyncracies, such as incomplete reporting, different academic practices and evaluations, and interpretations that vary over time. Metrics that do not clearly reflect a program's purposes and goals can seriously skew its performance. There is truth in the adage that "what you measure is what you get."
- Bibliometric analysis is important for program evaluation, but it must be conducted in conjunction with expert review; expert review will assist in placing the results of bibliometric analysis in the context of the current state of scientific knowledge.

REFERENCES

Cozzens, S. 2000. Higher education research assessment in the United States. Part 2 County Case-Study B in Valuing University Research: International Experiences in Monitoring and Evaluating Research Output and Outcomes, T. Turpin, S. Garrett-Jones, D. Aylward, G. Speak, R. Iredale, and S. Cozzens, eds. The Centre for Research Policy, University of Wollongong, Canberra [Online]. Available: http://www.dest.gov.au/archive/highered/respubs/value/susan cozintro.htm [accessed Jan. 22, 2003].

Cozzens, S.E. 2002. The Craft of Research Evaluation. Presentation at the Second Meeting on Review EPA's Research Grants Program, April 25, 2002, Washington, DC.

EPA (U.S. Environmental Protection Agency). 2001. ORD Strategic Plan. EPA/600/R/01/003. Office of Research and Development, U.S. Environmental Protection Agency, Washington, DC. [Online]. Available: http://www.epa.gov/ospinter/strtplan/documents/final.pdf [accessed Jan. 22, 2003].

GAO (U.S. General Accounting Office). 1997. The Government Performance and Results Act: 1997 Governmentwide Implementation Will be Uneven. GAO/GGD-97-109. Washington, DC: U.S. General Accounting Office.

Geisler, E. 2000. The Metrics of Science and Technology. Westport, CT: Quorum Books.

NRC (National Research Council). 1999. Evaluating Federal Research Programs: Research and the Government Performance and Results Act. Washington, DC: National Academy Press.

OMB (Office of Management and Budget). 2002. OMB Preliminary Investment Criteria for Basic Research. OMB Discussion Draft, February 2002.

OSTP/OMB (Office of Science Technology and Policy/Office of Management and Budget). 2002. FY 2004 Interagency Research and Development Priorities. Memorandum for the Heads of Executive Departments and Agencies, from John Marburger, Director, Office of Science and Technology Policy, and Mitchell Daniels, Director, Office of Management and Budget, The White House, Washington, DC. May 30, 2002.

Preuss, P.W. 2002. National Center for Environmental Research, History, Goals, and Operation of the STAR Program. Presentation at the First Meeting on Review EPA's Research Grants Program, March 18, 2002, Washington, DC.

5

Taking the Measure of STAR

This chapter contains the committee's evaluation of the Environmental Protection Agency (EPA) Science To Achieve Results (STAR) program. For its evaluation, the committee selected a set of metrics and recommends that EPA consider them as it adopts evaluative criteria for future evaluations of the STAR program. The evaluation is structured according to the guidelines that the president's Office of Management and Budget (OMB) issued for government agencies to use in assessing their research programs for FY 2004 (OSTP/OMB 2002). As described in Chapter 4, the OMB guidelines set forth three major criteria for evaluating research programs: quality, relevance, and performance. The committee considered that following the OMB guidelines would be most valuable to EPA as it strives to comply with the requirements of the Government Performance and Results Act (GPRA) of 1993.

Under each of the three OMB criteria, the committee has developed metrics related to processes and products of the STAR grants program. Specifically, the committee reviewed numerous documents, including materials from EPA, OMB, and the National Research Council (NRC) and materials provided by people in academe, to derive these metrics. The process metrics are used to evaluate the adequacy of the operation or procedures of the STAR program, and the product metrics are used to evaluate the outputs of the program, such as the number of publications or the influence or effect the program has had or may have. The committee evaluated the STAR fellowship program independently from the grants program because it is a small part of the STAR program and operates somewhat independently.

The committee was not able to evaluate some of the metrics completely, particularly those related to products, primarily because STAR has not been in operation long enough to produce a sufficient number of products to allow a complete evaluation. Other reasons for incomplete evaluations are that EPA has not collected sufficient information relevant to a metric and that some of the metrics are not intended to be addressed solely by the STAR program, but cover issues that are broader and that must be addressed by ORD and EPA. The set of metrics used by the committee in evaluating the STAR grants program is presented in Table 5-1. The goals and objectives of the fellowship program differ from the grants program, and it is evaluated more briefly at the end of this chapter. For each metric, this chapter summarizes the pertinent information provided to (or otherwise obtained by) the committee and then presents the committee's conclusions about how adequately the STAR program appears to be addressing it with respect to the available information—except that, as stated above, for some product metrics the STAR program is too young to have produced sufficient products to permit a complete evaluation.

RESEARCH PROGRAM

Quality

Evaluating research quality is extremely difficult. It cannot be measured with a simple metric, such as a thermometer or a yardstick, or by the number of reports or number of pages produced. As indicated in Chapter 4, even more sophisticated measures, such as the number of citations in the technical literature, need to be carefully interpreted.

The STAR program has tended to focus more on the quality of its process than on the quality of its products. That is understandable and appropriate. It is understandable because, being a relatively young program, STAR has had to focus on trying to get the process right and is only now beginning to accumulate a sufficient number of products to support a quality evaluation and because most of the external reviews of the program have tended to focus on process issues.

It is appropriate because, as indicated in Chapter 4, a high-quality process is generally a necessary condition for producing high-quality products. Frequent evaluations of the quality of the process will also provide an early warning of possible problems in the quality of the products. The causes of inadequate products usually lie in inadequacies of procedures that occurred many years previously.

TABLE 5-1 Metrics Used in the Committee's Evaluation of the STAR Grants Program

QUALITY
Process
Does the program have an effective process to ensure receipt of high-quality proposals for its grant awards?
Does the program have an effective process to ensure the selection of high-quality proposals?
Does the program have a mechanism for encouraging high-quality research?
Does the program have a clearly defined plan for regular, external reviews of its research quality, and has this plan been effectively carried out?
Product
Is the STAR program sponsoring high-quality research?
Has the program made significant contributions to advancing the state of the science in particular topics?
Do bibliometric and citation analyses demonstrate excellence in the program's research?
RELEVANCE
Process
Does the STAR portfolio support EPA's mission, GPRA goals, and ORD's strategic plans, research strategies, and multiyear plans?
Are the processes used to define the research initiatives that will be supported by the STAR program sufficient to target the topics of most important uncertainty, highest impact, or highest priority?
Does the program have a "clear plan for external reviews of the program's relevance" (OSTP/OMB 2002), and has this plan been effectively carried out?
Does the program have an effective process for identifying and communicating with the potential audiences and users of the research results?
Product
Is the STAR portfolio appropriately mixed between core and problem-driven research and between human health and ecologic research?
Does the program have a good plan for integrating and synthesizing results, and has this plan been carried out effectively?
Have the program's results been used in EPA, state, or international decision-support documents?

(Continued)

TABLE 5-1 *Continued*

RELEVANCE (cont.)
Product
Have the research results in one or more subjects significantly improved the scientific foundation for decision-making?
Can a link between STAR research and improved protection of human health and ecologic systems be identified?
PERFORMANCE
Process
Is the STAR budget appropriate to fulfill the program's mission?
Is the program effectively complementing ORD's other research efforts?
Is the program well balanced?
Does the program award grants expeditiously?
Does the program have a process to demonstrate the communication of individual grant results in the professional literature?
Is there a process in place for reviewing the performance of individual investigators and research centers?
Product
Is the program funding relevant research that otherwise would not be funded?
Does the program have a schedule for the products it intends to produce, and how well is it adhering to that schedule?
To what extent are site-specific studies designed to be replicated at other locations?

Although a high-quality process may be necessary for producing high-quality products, it is not sufficient. It is time for the STAR program to begin to implement product evaluations that will ensure that both its process and its products have the high quality that the nation needs to support an effective and efficient environmental-management program.

Process Metrics

Does the program have an effective process to ensure receipt of high-quality proposals for its grant awards?

EPA provided the committee, through presentations and interviews, with a substantial amount of information (summarized in Chapter 2) concerning how the STAR program attempts to elicit good proposals and how it evaluates those it receives.

There are two basic steps in establishing a good process for satisfying this metric. The first is to identify the topic to be addressed by the research in a way that stimulates good responses. The second is to advertise the availability of research support broadly in the research communities most qualified to undertake high-quality research on the desired topic.

With respect to the first step, EPA puts substantial effort into defining its research agenda, and the STAR program submits its proposed requests for applications (RFAs) to extensive review within the agency. That effort is intended to ensure that the RFAs are focused on the most important issues and that they define the research requirements properly. However, although substantial effort is devoted to the process in the agency, neither the research plans nor the proposed RFAs are externally peer-reviewed by subject-matter experts, except when the research is being supported jointly by another organization, in which case, representatives of the other organization participate in drafting and reviewing the RFA.

With respect to notifying potential researchers of the funding opportunities, the National Center for Environmental Research (NCER) makes a substantial effort to reach out to a broad scientific community and to recruit the most capable scientists. NCER disseminates its RFAs widely through its Web site, the *Federal Register*, announcements at professional meetings, and e-mail distributions to individuals or institutions that sign up on the STAR Web site. When the desired research falls outside EPA's traditional research fields and may therefore involve scientists that are not already tied into the agency's research program, STAR solicits the help of other agencies that traditionally work with these scientists to ensure that they are aware of the funding opportunities.

The committee concludes that the processes established by the STAR program compare favorably with and in many cases substantially exceed those established by other research-supporting organizations. Subjecting the research plans and draft RFAs to independent peer review would strengthen the process even more, but the committee recognizes that doing so might reduce the agency's current flexibility to respond quickly to new research needs and might unduly delay the process of issuing RFAs. However, EPA should consider using external peer reviewers for RFAs when they do not have the in-house expertise.

Does the program have an effective process to ensure the selection of high-quality proposals?

The process for selecting proposals was described to the committee by the person responsible for managing it and by several STAR project officers, and some committee members had participated in the process previously.

The STAR program has established a rigorous peer-review process to evaluate the quality of proposals. Such peer review is the foundation on which excellence is achieved in all research programs, such as those of the National Institutes of Health (NIH) and the National Science Foundation (NSF). The agency has taken strong steps to ensure that this process does not suffer from conflicts of interest and is independent. The program's procedures provide for a firewall that shields the peer-review process from any influence or potential conflicts of the project officers and staff who oversee the individual investigator, fellowship, and center awards. For instance, project officers can provide the names of potential reviewers to the Science Review Administrators (SRAs), also known as peer-review officers, but it is the sole responsibility of the SRAs to select reviewers and to make reviewer assignments. Project officers may attend peer-review meetings as observers but may not provide any comments that would affect peer review (Bryan 2002).

The program selects peer reviewers from a large number of sources, including people who have served on previous panels, keyword searches of databases (such as, Community of Science, the National Library of Medicine's PUBMED, and faculty listings), keyword searches of NCER's peer-review panelist information system, input from project officers and program-office scientists, and lists of the attendees of pertinent technical conferences (Bryan 2002). The agency is unusual in that it pays the members of its peer-review panels a daily stipend, presumably to provide additional encouragement for experts to participate. Although lower than the consulting fees that such experts might earn in the private sector, the stipends do reduce the financial disincentives associated with serving on the panels.

The committee received some comments suggesting that in the program's early days, some members of the peer-review panels might not always have had the necessary qualifications to be effective members, but this problem appears to have disappeared as the program has matured.

The committee heard no suggestion that the process of selecting peer reviewers was influenced by conflicts of interest.

The final selection of proposals to be funded is influenced by a "relevance review" that is carried out by agency staff. That review is limited to proposals that have been rated as "excellent" or "very good" by the quality peer-review panel.

The committee concludes that, given EPA's desire to avoid appearances of conflicts of interest by completely separating the selection of peer-review panel members from the influence of project officers, the agency has established a rigorous, independent peer-review process for selecting the highest-quality proposals.

Does the program have a mechanism for encouraging high-quality research?

To gain a better understanding of how the STAR program encourages high-quality research, the committee received briefings from NCER staff on its processes, reviewed material that was publicly available, attended several progress-review meetings, and discussed the program's procedures with STAR grant recipients.

STAR has implemented several mechanisms for encouraging high-quality research by its investigators. Investigators are required to submit annual progress reports that describe the research being undertaken and its progress. The progress reports are reviewed by STAR project officers, and summaries are posted on the NCER Web site.

In addition, the STAR program sponsors progress review workshops on research topics. Principal investigators of all STAR grants receiving support pertinent to the topic are expected to attend. The meetings are also attended by other EPA Office of Research and Development (ORD) staff conducting relevant research; all meetings are open to the public. At the meetings, principal investigators must present their research progress to their colleagues and EPA staff, opening it to peer review. The meetings also provide an opportunity for researchers to share ideas and coordinate research efforts. Some of the meetings have apparently been much more successful than others in accomplishing their objectives.

The committee concludes that EPA has established procedures for reviewing the quality of research in progress that in several ways exceed those adopted by most other research-supporting organizations. Those procedures could be enhanced by ensuring that progress-review workshops were held at the most expeditious times and were run most efficiently to stimulate peer review and collaboration. The program should review the success of its past progress-review meetings and organize its future meetings to emulate the ones that were most successful in accomplishing their objectives.

Does the program have a clearly defined plan for regular, external reviews of its research quality, and has this plan been effectively carried out?

As mentioned in Chapter 2, numerous reviews have been done on the processes and operation of the STAR grants program, and not on the products of the grants (EPA/BOSC 1998; EPA 2000; EPA/NSF 2000; EPA/SAB/BOSC 2000; GAO 2000; EPA/SAB 2001; EPA 2002a). The majority of the reviews have been conducted by EPA's Science Advisory Board (SAB) or Board of Scientific Counselors (BOSC).

The committee is concerned that too many and too frequent reviews of the STAR program have the potential to be damaging in that they may divert necessary financial and personnel resources from the program. The committee believes strongly that because of the nature of research, which takes a considerable amount of time and many projects to advance the state of knowledge, too frequent reviews of STAR add little value to the understanding of the operation and results of the program.

The committee recommends that the STAR program establish a schedule of product reviews at the appropriate level (as discussed in Chapter 4). By establishing such a schedule, the program may protect itself from the apparent excess of external reviews that have been imposed on it in the past.

Product Metrics

Is the STAR program sponsoring high-quality research?

The committee was presented with some anecdotes concerning the quality of the research being sponsored by the STAR program, but not with any systematic reviews or evidence concerning the quality of the program's products.

Evaluating the quality of research products is very difficult, involving substantial judgment on the part of scientists who have expertise in the research topic being reviewed. As indicated in Chapter 4, the committee considers that the most effective method for evaluating research is the use of independent expert review committees focusing on specific topics (a level 2 review). The committee recommended that the STAR program establish a schedule for such reviews (see Chapter 4).

Lacking product evaluations, the committee reviewed the backgrounds and accomplishments of a sampling of the STAR principal investigators. Although a rigorous sampling procedure was not conducted and thus the results of the review are only indicative and not definitive, this review indicated that the STAR program is funding many scientists who have outstanding research credentials. The scientists have impressive track records and are leaders in their fields. Many are editors of journals or officers in societies and have received awards of distinction. Some were attracted to the STAR program from fields outside EPA's mission, so it can be said that the program has been successful in attracting the best and the brightest. Many of the investigators, however, have long been active in the relevant fields, and the STAR program has enabled them to continue to make contributions. The investigator mix also included young investigators who will be the leaders of the future.

The committee notes that EPA's rate of funding of individual investigator and center awards tends to be lower than that of other federal grants programs, such as those sponsored by NSF and NIH; this reflects the competitiveness of the program. As indicated in Chapter 2, data from FY 1999-2001 indicate that EPA funds an average of 10-15% of the proposals it receives. In contrast, as indicated in Chapter 3, agencies like NIH and NSF strive to fund at least about 25-30% of the proposals received. STAR is able to fund only about 60% of proposals rated as "excellent" or "very good" by its independent quality peer-review panels. It funds no proposals that receive a lower ranking ("good" or lower).

On the basis of the STAR program's process for awarding grants, the quality of the people and institutions being funded by the STAR program, and the highly competitive nature of its awards, the committee is confident that the products of these grants will be of the highest quality.

Has the program made significant contributions to advancing the state of the science in particular topics?

Although the STAR program does not systematically identify the significant contributions it makes to filling important gaps in the state of the science in particular topics, examples were presented to the committee by EPA staff and by investigators supported by STAR funding. Some of the committee members were also familiar with the advances that STAR-funded research was making to some particular topics.

The planning process that leads to the preparation of RFAs is to a large extent focused on ensuring that STAR grants will fill information gaps in topics of greatest interest to the agency. That focus is maintained throughout the research process.

The committee was presented with several examples of STAR-supported research efforts that had made significant contributions to scientific understanding in particular topics (see Boxes 5-1, 5-2, and 5-3). For example, STAR-sponsored research in endocrine disruptors, particulate matter, and ecologic assessment has resulted in peer-reviewed groups of publications of immediate interest in understanding causes of, exposures to, and effects of environmental pollution.

To determine whether STAR research has filled a critical knowledge gap or otherwise strengthened and improved the foundation for decision making, it would be useful to assess the state of the science in a particular issue before STAR-funded projects are completed and then synthesize the results of the research after the projects are completed. Such assessments would help EPA to target RFAs at the front end, as well as to analyze net results at the back end. A particular issue could be assessed by a panel convened by ORD, by STAR, or by others in the field (such as the Ecological Society of America, or the Society of Environmental Toxicology and Chemistry). There have been several successful examples of state-of-the-science documents, including the particulate-matter (PM) reports produced by the NRC's Committee on Research Priorities for Airborne Particulate Matter (NRC 1998, 1999b, 2001). In general, better integration of research results by STAR and ORD and the state-of-the-science assessments mentioned above should provide most of the information necessary to report on this metric.

The committee recognizes of course that this metric involves substantial subjective judgment and that it is often difficult to identify the effect of any particular set of research results. The judgments implicit in this metric can probably be best rendered by the use of expert review, as suggested in Chapter 4.

The committee concludes that STAR-supported research is making significant contributions to advancing the state of the science in many of the topics that it is addressing. The committee suggests that the program undertake a more systematic effort to identify the contributions and the success in filling the knowledge gaps identified in the research planning process by preparing research synthesis reports when research is completed, as recommended elsewhere in this chapter.

BOX 5-1 Results of STAR Endocrine-Disruptors Program

- Determined that exposure to high concentrations of polybrominated biphenyls prenatally and in breast milk may affect puberty in girls (Blanck et al. 2000).
- Discovered a new (third) estrogen receptor in vertebrates and demonstrated that estrogens and xenoestrogens can act on cells at the membrane level (Hawkins et al. 2000).
- Developed and refined an in vivo model using medaka to identify endocrine disrupting chemicals (Cooke and Hinton 1999).
- Identified androgenic compounds (male-hormone mimics) in paper-mill effluent by using a screening assay in fish (Jenkins et al. 2001).
- Developed integrated array of computational tools undergoing validation by EPA's Office of Pollution Prevention and Toxic Substances (OPPTS) for setting priorities for science and technology programs (Xing et al. 1999).
- Determined concentrations of phytoestrogens in human amniotic fluid and effects of exposure to them in animal models (Hughes et al. 2001).

Source: Adapted from P. Preuss, EPA, presentation to National Research Council committee, March 18, 2002.

Do bibliometric and citation analyses demonstrate excellence in the program's research?

Although EPA encourages its grantees to provide the STAR program with information on the articles and other publications that stem from the research it supports, the program has no mechanism for monitoring such publications after a grant is completed or for conducting bibliometric and citation analyses that would demonstrate the influence of the research on other work being conducted. The committee did sponsor an ad hoc bibliometric analysis to gain a better understanding of the value of such an approach (IISCO, Atlanta, GA, unpublished material, 2002) (see Chapter 4). As indicated in Table 4-1, the bibliometric analysis conducted for the committee indicated that the citation rate of publications that result from STAR-supported research is similar to that of other research publications in the topics for which the analyses were undertaken.

As stated in Chapter 4, the committee considers that bibliometric and citation analyses are important quantitative metrics for gauging the quality of research but cautions that these types of analyses have many limitations.

BOX 5-2 Sampling of Results from the STAR Particulate-Matter Program

- **Are diabetics more susceptible to the health effects of airborne particles?** This study examined whether diabetes modifies the effects of PM. Researchers studied the association of PM_{10} with hospital admissions for heart and lung disease in persons with and without diabetes as a comorbidity. Using Medicare data for Cook County, Illinois, the investigators found that a 10-ug/m^3 increase in PM_{10} was associated with a 2.01% increase in admissions for heart disease with diabetes but only a 0.94% increase in persons without diabetes. Similar effect modification was not seen for lung diseases. When analyzing by age, researchers found twice the PM_{10}-associated risk for heart disease in diabetics as in nondiabetics in both age groups examined. Investigators concluded that people with diabetes are a more susceptible population (Zanobetti and Schwartz 2001).
- **Long-term effects of PM exposure include lung cancer.** A major epidemiology study determined that long-term exposure to combustion-related fine-particle air pollution is an important environmental risk factor for cardiopulmonary and lung-cancer mortality. The magnitude of lung-cancer mortality associated with fine-particle exposure has been equated to that from exposure to environmental tobacco smoke. The investigators linked data on 500,000 adults in the American Cancer Society's prospective mortality study with air-pollution data for metropolitan areas throughout the United States (Pope et al. 2002).
- **Early findings on biologic mechanisms associated with exposure to concentrated air particles.** Inhalation of concentrated urban air particles by rats results in modest pulmonary inflammation in normal animals and model populations in most experiments. Increases in circulating neutrophils as evidence of systemic inflammation were variable in these experiments. However, when circulating neutrophils were increased, there was not a measurable increase in circulating cytokines. In studies of rats with induced pulmonary inflammation, exposure to concentrated urban air particles resulted in the deaths of some animals with electrocardiographic evidence of sympathetic nervous system enhancement and arrhythmia. This research contributed to the beginnings of our understanding of the biologic mechanisms by which inhaled ambient particles cause health effects (Clarke et al. 1999; Lovett et al. 1999).
- **Chemical composition of atmospheric ultrafine particles.** Measurements of ultrafine particle mass concentration in seven southern California cities show that the chemical composition of these ultrafine particle samples averages 50% organic compounds, 14% trace-metal oxides, 8.7% elemental carbon, 8.2% sulfate, and 6.8% nitrate. A source emissions inventory constructed for the

(Continued)

BOX 5-2 *Continued*

South Coast Air Basin that surrounds Los Angeles shows a primary ultrafine particle emission rate of 13 tons/day. Those ultrafine emissions arise principally from mobile and stationary fuel-combustion sources and are estimated to consist of 65% organic compounds (Cass et al. 2000).

Source: S. Katz, EPA, personal commun., November 27, 2002.

For instance, average citation rates differ among topics and disciplines. It is also important to consider the prestige of the journals in which articles are published. Such analyses must be interpreted by experts familiar with the particular topics in question.

The committee also has concerns about the utility of bibliometric analysis to STAR, in as much as the program has not been in existence long enough to make bibliometric analysis an effective metric for research funded beyond the initial years of the program. Similar sentiments are echoed in the BOSC review of the STAR program (EPA/SAB/BOSC 2000) in which BOSC, in recommending that citations be used as a metric, cautioned that "4 to 6 years must pass between the completion of the STAR-funded research and the use of citations to judge success."

The committee recommends that EPA undertake bibliometric and citation analyses not as independent evaluations of research quality but in conjunction with expert review. The expert review will assist in placing the results of the bibliometric analysis in the context of the current state of scientific knowledge.

Relevance

In its discussion of relevance, OMB asks (OSTP/OMB 2002), "Does the agency's research address subjects in which new understanding could be important in fulfilling the agency's mission?" In the committee's view, mission relevance requires that STAR research improve the knowledge base required to identify environmental issues and make sound environmental decisions. The current STAR portfolio includes both core research (research that provides an understanding of the structure and function of environmental systems, the effects of human perturbations on those systems, and the resulting effects on human health and quality of life) and problem-driven research (research that focuses more specifically on questions related

BOX 5-3 Sampling of Achievements of STAR Ecosystem-Protection Program

- Developed estuarine index of biotic integrity that has shown broad applicability within the southern New England ecoregion. This index could be a valuable monitoring tool to assess the recovery of ecosystem function after eutrophication remediation (Deegan et al. 1997).
- Developed a dynamic, ecologic, economically linked model to evaluate the driving forces and ecologic consequences of land-use change. This model was used to demonstrate how changes in zoning would affect water quality in Calvert County, Maryland (Voinov et al. 1999).
- Developed and tested an integrated ecologic-assessment and decision-support tool for the Lake Erie ecosystem. The method was designed to assist managers and stakeholders involved in the Lake Erie Lakewide Management Plan (LaMP) and other Lake Erie management processes to define objectives and evaluate tradeoffs and risks associated with future uses (Locci and Koonce 1999).
- Assessed the impact of SO_2 and NO_2 emission reductions on precipitation and air quality by comparing emissions before and after the Clean Air Act Amendments. Results show a significant reduction in SO_2 emissions for most states except Texas, North Carolina, Illinois, Florida, and Alabama. However, only two states show a reduction in NO_x (Butler et al. 2001).
- Developed a reliable method for the simultaneous measurement of concentrations of viruses, bacteria, and protozoans in water using disposable hollow ultrafilters; this method allowed for *Cryptosporidium* oocyst recoveries of about 60-80%. The new method was published, making it available to the scientific community, EPA, and the general public for use in place of other, less accurate methods (Juliano and Sobsey 1997).
- Developed and validated a new approach for using satellite radar imagery to monitor the spatial and temporal patterns of flooding and drying in the wetland ecosystems of south Florida. This technology has been used to monitor the effects of human activities (such as the construction of roads and the operation of water-control structures) on natural hydroperiods in south Florida wetlands (Kasischke 2000).

Source: B. Levinson, EPA, personal commun., October 16, 2002.

to upcoming regulatory decisions). Although the core research is farther removed from the agency's regulatory efforts, it can provide new understanding important to the agency for fulfilling its mission.

Although the agency is the primary audience for the STAR program,

EPA's mission to safeguard the environment includes not only the performance of its own regulatory tasks but also providing the scientific foundation for decisions by other entities. In addition to its own efforts, EPA research aids environmental decision making by state, local, tribal, and international government agencies, other federal agencies, and the public. Accordingly, it is relevant for ORD and STAR research portfolios to incorporate research that will promote sound environmental decision making by all those users.

The committee believes that, first, the STAR portfolio should be related to EPA's mission and focus on EPA's highest-priority environmental science and engineering needs. Second, the body of knowledge provided by STAR research should contribute to identifiable progress toward practical outcomes. Third, a smaller but significant portion of STAR research portfolio should explore potential environmental problems. The metrics addressing the relevance of the STAR program, discussed below, touch on each of those general intents.

Process Metrics

Does the STAR portfolio support EPA's mission, GPRA goals, and ORD's strategic plan, research strategies, and multiyear plans?

The committee heard presentations from several EPA representatives regarding the agency's research planning process, discussed this process with other NCER staff, and reviewed information relevant to this metric that was available on EPA's Web site.

ORD invests a substantial amount of effort in the research planning process to ensure that the STAR portfolio will be relevant to EPA's mission and enhance ORD's ability to meet its strategic goals. For instance, STAR has adopted a planning process in which the development of the STAR portfolio evolves from EPA's strategic plan and GPRA goals, ORD's strategic plan, research strategies, and multiyear plans.[1] The RFAs, which are designed to carry out the research plans, are circulated throughout the rest of the agency to ensure that they are consistent with them.

---

[1]In general, the distinction between these is that the EPA and ORD strategic plans define the universe of mission-related activities that ORD should target; the ORD research strategies define which issues are important and frame the scientific questions associated with them; and the ORD multiyear plans apportion the research between intramural and extramural (STAR) programs.

However, research strategies and research plans do not necessarily exist for all the topics supported by STAR grants; and even when they do, they are often internal working documents and are not available for public review. The ecologic indicators (EPA 1998a) and airborne particulate matter (EPA 1999) research strategies were available, and comparison of the strategies with the relevant RFAs shows that the STAR program has been designed to answer the research questions posed by the strategies. The endocrine-disruptors research plan (EPA 1998b) tracks closely with the research questions addressed in the RFAs.

In addition, review of all the RFAs issued during FY 1999, 2000, and 2001 shows that the topics addressed in the RFAs are mission-relevant and related to the EPA and ORD strategic plans.

Therefore, on the basis of the information available, the committee concludes that the STAR portfolio does support EPA's mission and GPRA goals and ORD's strategic plans, research strategies, and multiyear plans.

Are the processes used to define the research initiatives that will be supported by the STAR program sufficient to target the topics of most important uncertainty, highest impact, or highest priority?

The processes used to define STAR research initiatives are entrenched in the agency's extensive research planning. Most of the planning efforts are internal and were not available for review by the committee.

The process used to develop RFAs is designed to focus STAR research on gaps in knowledge related to EPA's mission, its high-priority research needs, and subjects with the greatest uncertainty and potential impact. However, only EPA staff have a substantial input into this process. In some circumstances, the expertise of the agency personnel may not always be sufficient to select among the various options for deciding the highest-priority research for funding, particularly when the issue being addressed by the research falls outside EPA's regulatory purview.

One activity that would assist in the identification of highest-priority research would be a survey of the state of the science in potential research topics before the designing of RFAs. In some cases, that would require additional resources, for example, to convene a panel charged with reviewing the state of the science and suggesting topics in which new research may have the greatest impact on EPA's mission (see also the discussion of "relevance-review" panels in NRC 1999a). In others, particularly those for which detailed research plans have been developed, additional advice from selected outside experts may suffice.

The committee's brief review of the RFAs for research in ecologic indicators, endocrine disruptors, and particulate matter showed increasingly useful focus on critical issues and questions as the programs matured. In fact, an EPA program officer commented that RFAs have become more specific as the agency has learned that a field is advanced further when research is more specifically targeted (E. Francis, EPA, Washington, D.C., personal commun., August 6, 2002).

Although the committee concludes, on the basis of available information, that the STAR program has been increasingly successful in identifying and defining research efforts that are of highest priority, it recommends that the agency explore methods for improvement, particularly when research is not directly related to the agency's regulatory programs. A potential method of incorporating a relevance analysis into grant applications was offered by the EPA SAB (EPA/SAB 2001). In this review of the STAR Water and Watersheds program, the SAB suggested that grant applications incorporate a number of questions that would require the researchers to evaluate how the proposed research would inform future watershed management decisions, provide increased understanding of the structure and function of watersheds, and relate to the priorities listed in ORD's strategic plan.

Does the program have a "clear plan for external reviews of the program's relevance" (OSTP/OMB 2002), and has this plan been effectively carried out?

The committee examined copies of all the reviews of the STAR program it received from EPA. The report by the General Accounting Office (GAO 2000) commented explicitly on the relevance of the STAR program to the EPA program offices, but this was not a "planned review."

The research planning processes and the agency's budgeting process provide an element of such reviews in that the program's primary audience, EPA's program offices, has a substantial opportunity to comment on the relevance of the program's research and to influence its funding. However, those activities may not fall into the category of "external reviews" as defined by OMB, and they do not include audiences outside the agency.

The committee concludes that the program has no established plan for external reviews of its relevance and recommends that, after the program has adequately identified its potential audiences as recommended below, it institute such a plan as discussed in Chapter 4.

Does the program have an effective process for identifying and communicating with the potential audiences and users of the research results?

From EPA presentations to the committee and interviews with NCER staff, the committee obtained substantial information about the STAR program efforts to be responsive to the needs of the scientific community, EPA program offices, and other potential audiences. Committee members also attended some of the outreach programs that STAR sponsored. However, the committee received little information about efforts to identify other potential users and to ensure that research results are relevant to them.

In response to questions raised in a report to ORD's BOSC, NCER indicated that it "believes that its audiences include not only its own organization, ORD, and EPA's regional and program offices, but also other agencies and professional societies, and the outside world of Capitol Hill, the scientific community, and the public" (EPA 2002a).

Although the STAR program does not have a process for identifying its potential audiences, EPA seems to recognize the need to disseminate research results to a wider audience. Indeed, as the STAR program has matured, it has devoted increasing effort to finding effective means of communicating its research results to potential audiences. It provides a considerable amount of information about its projects on the NCER Web site, and the quality of the reports it makes available there has improved. The program has also experimented with some unique mechanisms for communication. For instance, the STAR program and EPA Region III supported the Mid-Atlantic Integrated Assessment project, which was designed to acquaint state and local agencies and environmental and community organizations in the mid-Atlantic region with the results of STAR-supported research efforts and to assess the relevance of the results to their needs (Bradley 2002). The STAR program also sponsored a workshop with Region I (New England) that was intended to bring STAR research results to state, federal, and tribal environmental programs in New England.

However, the program has not yet effectively developed a strategy for communicating to its wider user community; most of the emphasis has been on the scientific community and the EPA program offices. Any user who lacks the sophistication or incentive to visit EPA's Web site and wade through individual grant reports or research publications is unlikely to benefit directly from the program.

One difficulty that the program faces in identifying and communicating with its audiences is that much STAR-sponsored research is designed to complement other research efforts undertaken by ORD. In many cases, the relevant question is whether the entire EPA research effort, not just the part supported by STAR, is satisfying the user community. The user community is likely to have little knowledge of and to be largely uninterested in who produced the information. The user community's primary interest is that the information be available and dependable. Indeed, the EPA program office staff, especially those who do not routinely review the scientific literature, are generally not aware of STAR results. However, since this is the primary audience for the STAR program, it is clearly in the program's best interest to find ways of making this audience more aware of its value.

The committee understands that serving and effectively communicating to a diverse audience is difficult for a research organization, and it commends the STAR program for the innovative approaches it has developed to improve its communication efforts. However, an effective process for communicating the results of the STAR grants with its wider user community remains among the most important improvements that can be made in the program. The first step in accomplishing that is to define the audiences better. The committee recommends that the program, in association with the rest of ORD, then develop aggressive plans to disseminate STAR results to the audiences.

Product Metrics

Is the STAR portfolio appropriately mixed between core and problem-driven research and between human health and ecologic research?

In presentations to the committee, STAR representatives often mentioned the desire to maintain a balance between core and problem-driven research as called for in ORD's strategic plan and between research efforts focused on human health and ecologic assessment.

Assessing whether a research program has maintained such balances is difficult and subjective. It is often difficult to determine how a particular RFA should be classified as to whether it is core or problem-driven. If it addresses both human health and ecologic effects, how should it be classified, particularly if much of what is termed ecologic research actually elucidates the processes by which environmental stressors affect both humans and other ecosystem attributes?

The committee did not attempt to conduct a comprehensive review of all of STAR's individual grants and center research plans to determine whether such balances existed. However, a brief review of the RFAs for

FY 1999, 2000, and 2001 by the committee indicates that there was a rough balance between core and problem-driven research. In addition, the STAR portfolio has maintained a rough balance between human health and ecologic research, althoughdata for FY 1998-2002 indicate a recent tendency to tip the budget more toward human health (J. Puzak, EPA, Washington, D.C., personal commun., April 2, 2002).

Determining whether there is a balance is difficult enough. Assessing whether a given balance is "appropriate" is largely subjective, particularly because an effort to maintain a balance may be in conflict with the program's goal to fund the highest-priority research. Should the program's limited funds be used for research projects that will help to maintain a balance even if the projects are not of the "highest priority"?

The answer to that question is probably a qualified yes. For instance, the tight deadlines often imposed on EPA by Congress can force the agency to focus all its attention on what needs to be done tomorrow, but an effective environmental-management program requires a view of the future and the problems that it may hold. Similarly, some types of problems may, because of congressional mandates or other pressures, be more immediate, but few problems disappear, and it would be inappropriate for a research program to vacillate from one problem to another.

Given all the difficulties of making such an assessment, the committee nevertheless concludes that the STAR program has generally maintained appropriate balances between core and problem-driven research and between human health and ecologic research, and it encourages the program to maintain such balances in the future.

Does the program have a good plan for integrating and synthesizing results, and has this plan been carried out effectively?

Several of the EPA presentations to the committee addressed plans for integrating and synthesizing research results. Committee members obtained additional information from interviews with NCER staff.

As indicated in Chapter 2, the STAR program produces two types of reports, *STAR Research Capsules* and *STAR Integrated Topical Searches*, that provide summaries of individual research efforts addressing a particular topic. The program has also attempted to prepare synthesis reports that provide an integrated summary of the research that has been carried out on a particular topic. However, the program has apparently had difficulties with those efforts; no reports have been publicly released.

The committee believes that such synthesis reports can be very valuable, both for presenting a coherent summary of what is known about a particular topic and for targeting gaps in knowledge. The appropriate type of synthesis document will, however, depend on the intended audience. In many cases, because the STAR projects represent only a part of the important research being done on the topic, such synthesis reports would need to include the research undertaken by other ORD components and perhaps other research centers. To reduce possible concerns about the objectivity or independence of these reports, they should not be prepared by EPA staff. The committee believes that supporting the preparation of synthesis reports is an important improvement that should be made in the STAR program.

Have the program's results been used in EPA, state, or international decision-support documents?

The STAR program has not been in existence long enough to be able to document the extent to which its research results are being used in supporting new regulations and other environmental-management decisions. Even for the projects that have been completed, there is often a substantial delay between the time that the research is completed and the agency's decision to undertake rule-making or other actions to address the issues being studied.

The committee received no information about any efforts by or plans for the STAR program to collect information on the extent to which its results are being used in decision-support documents. Because that is one of the primary purposes of the STAR program, it is reasonable to expect the program to collect information about its success in fulfilling it, at least with respect to the agency's own decisions. Attempting to monitor the use of STAR results by other agencies and organizations would be more difficult.

The committee recommends that the STAR program consider developing a mechanism for monitoring the use of its research results in criteria and other documents that support agency environmental-management decisions. For instance, EPA should attempt to keep track of the number of times STAR-funded research projects are cited in decision-support documents. However, an independent expert review of such an analysis would be necessary to assess the value or impact of this research in decision making.

Have the research results in one or more subjects significantly improved the scientific foundation for decision making?

As indicated earlier, the committee received anecdotal evidence that identified topics in which STAR research had resulted in groups of peer-reviewed publications of immediate use in understanding causes, exposures, and effects of environmental pollution, such as those cited for STAR in endocrine disruptors, PM, and ecologic assessment (see Boxes 5-1, 5-2, and 5-3). Additional time may be needed to view the full influence of the research.

This metric goes to the heart of relevance for a mission-related agency such as EPA. Research that improves the scientific foundation for environmental decision making incorporates both core and problem-driven research, and it incorporates the research performed not only by STAR but by ORD as a whole. Accordingly, this metric looks at each of the major subjects addressed by the STAR program and asks whether the results of the STAR-funded research—in combination with research performed elsewhere—filled a critical knowledge gap, provided an important decision-support tool, or otherwise improved the ability to target and manage an environmental effect. For some mature issues, this metric could include evidence that research already has improved risk-assessment and risk-management decisions. For example, the BOSC review of the STAR program (EPA/SAB/BOSC 2000) suggested tracking evidence that STAR research has enabled the agency to manage hazards that had not previously been identified as conferring significant risks, implement more cost-effective remedies for known hazards, or reduce the stringency of regulations for hazards that are found to be less serious than previously thought.

The committee concludes, on the basis of the data that it was provided (see Boxes 5-1, 5-2, and 5-3), that some STAR research efforts have already significantly improved the scientific foundation for decision making. However, the committee recommends that the program initiate a more coherent plan for identifying such instances.

Can a link between STAR research and improved protection of human health and ecologic systems be identified?

This type of question, focused on the ultimate purpose of government programs, is being asked of all their activities. The committee recognizes that establishing a link between research and protection of (or improvement in) human health or ecologic conditions is extraordinarily difficult, because temporal lags are long, with different types of research affecting human health and ecologic conditions in very different time frames, and because monitoring of human health and ecologic conditions is spotty. However, the ultimate mission of EPA is to "protect human health and to safeguard the natural environment—air, water, and land—upon which life depends."

It is appropriate, therefore, to assess this metric periodically as a touchstone for STAR in the context of ORD research, even while recognizing that it will be difficult to document successes.

Performance

In its guidelines regarding program performance, OMB emphasizes metrics that show success in achieving "identifiable results" in keeping with a schedule of "multi-year R&D objectives with annual performance outputs and milestones" (OSTP/OMB 2002). Performance, however, also includes how a program achieves it goals. Is it an open, transparent process devoid of conflicts of interest and biases? Are procedures clearly established and predictable? Is it well integrated into the agency's other research efforts, and does it expeditiously communicate its results to potential users?

Some of those questions were addressed in connection with quality and relevance metrics. Others are addressed here.

Process Metrics

Is the STAR budget appropriate to fulfill the program's mission?

The committee reviewed a substantial amount of information about the details of the STAR budget but less about the appropriateness of the available funding to meet the program's mission.

After growing rapidly during the program's early years, the STAR budget has remained relatively constant, around $100 million per year for the last several years. Research costs have increased somewhat during this period, resulting in a decline in the amount of research that can be supported by the program. The lack of growth in the STAR budget, however, reflects the lack of growth in EPA's total research budget. In fact, the proportion of the research budget allocated to the STAR program has increased slightly and now stands at about 18% of the ORD budget.

The STAR program has leveraged its resources by forming partnerships with other agencies that have similar research interests. EPA reports that STAR partnerships with other federal and private organizations leveraged funds in such a way as to permit 35% more grants to be funded than would have been possible with EPA resources alone (P. Preuss, EPA, Washington, D.C., personal commun., March 19, 2002). About half the STAR annual announcements are funded jointly with other federal agencies (P. Preuss,

EPA, Washington, D.C., personal commun., March 18, 2002). (See Chapter 2 for additional information on joint funding.)

EPA's joint funding with other federal agencies provides several obvious advantages, including the funding of a larger amount of research that is tied directly to the agency's mission and the pursuit of a broader research portfolio. One example of those benefits is the 2001 RFA on epidemiologic studies on the effects of exposures to endocrine disruptors that was issued jointly by EPA, National Cancer Institute (NCI), and National Institute for Occupational Safety and Health (NIOSH). EPA funded the studies focused on specific endocrine-disrupting chemicals of interest (such as phthalates and polybrominated diphenyl ether), NCI funded the cancer studies, and NIOSH funded the occupational studies (E. Francis, EPA, Washington, D.C., personal commun., August 6, 2002). EPA commented that its greatest challenge with the joint funding efforts is to ensure that the RFAs fully meet EPA's mission and research needs and that the agency does not simply fund more but less-relevant research (P. Preuss, EPA, Washington, D.C., personal commun., June 6, 2002).

The committee concludes that the STAR program budget—which funds a mix of individual investigator awards, centers, and fellowships—represents an appropriate proportion of the ORD portfolio because it provides research capabilities that EPA may not have in-house or that complement the agency's intramural research. However, the committee is concerned that the STAR budget has not kept pace with general inflation or the rising costs of research and education. The lack of growth in EPA's total research budget may be preventing the agency from establishing the firm information base it needs to support its environmental-management efforts. In the future, an expert committee may want to tackle the question of whether the budget of the STAR program, and more generally the entire R&D apparatus of EPA, combined with other federal R&D expenditures, is sufficient to provide the level of research and technology necessary to provide the resources and knowledge base for the future.

Is the program effectively complementing ORD's other research efforts?

The committee heard several presentations about the agency's research-planning processes, and it interviewed STAR project officers responsible for the PM, endocrine-disruptors, and ecologic-indicators programs about their efforts to ensure that STAR supported research is complementary to other research being sponsored by EPA.

EPA has adopted a multilevel, comprehensive research-planning process that has a major goal of ensuring that the agency's diverse research efforts complement one another. Topics assigned to the STAR program are ones that ORD's internal research capabilities are not able to address effectively in a timely manner or for which independent analyses are desired. The STAR program draws on experts outside the agency to address mission-related research questions. EPA is thus able to develop and foster relationships with researchers to address agency needs without greatly expanding in-house staff. The agency is able to use those researchers when necessary to respond to issues that are not within the expertise of in-house scientists or when required to address emerging problems.

The discussions with the STAR project officers and other EPA officials within ORD indicated that STAR-funded research complements and enhances EPA's in-house research. For instance, for the endocrine-disruptors program, STAR funding is used to fund research in topics not being addressed in-house. The STAR program has funded exposure assessment and epidemiologic studies on endocrine disruptors—subjects on which in-house laboratories have not focused. Similarly, for the ecologic-indicators program the majority of research on the development of indicators is funded through the STAR program, and in-house research is focused on the "proof of concept" or "implementation" of the indicators in the field (B. Levinson, EPA, Washington, D.C., personal commun., August 5, 2002).

The STAR program further attempts to ensure and promote complementarity through its progress-review workshops, which include all STAR-supported principal investigators and representatives of all the other ORD research facilities working on a common topic.

The committee concludes that the STAR program is well coordinated with the rest of the agency's research efforts and is a necessary asset of these efforts. The STAR program allows EPA to access resources that are not available in-house, thereby substantially expanding the scientific capabilities available to assist the agency in formulating its environmental-management initiatives.

Is the program well balanced?

The committee heard presentations from several NCER officials about the balance that the STAR program maintains between different types of research, including core and problem-driven, ecologic and health, and center and individual-investigator awards.

As indicated in Table 2-1, the STAR program maintains a diverse research portfolio. Much of the research is focused on particular issues of immediate interest to the agency, but some addresses more basic, "core," or possible future environmental issues. Much of the research addresses human health issues, but this is balanced by ecologic and socioeconomic topics. Some research is undertaken by individual investigators, but multidisciplinary research centers are increasingly important.

The program appears to face a tension between maintaining a balance among different subjects of research and addressing the most important issues that the rest of the agency is attempting to deal with. To some extent, those distinctions are not as meaningful as they might appear. For instance, much of what is termed ecologic research actually elucidates the processes by which environmental stressors affect both humans and other ecosystem attributes. Many of the issues of balance were addressed in the discussion under the metric *Is the STAR portfolio appropriately mixed between core and problem-driven research and between human health and ecologic research?* in the section on relevance, discussed earlier.

The question of whether the program should depend primarily on individual investigator awards or continue to increase the proportion of the budget allocated to supporting research centers appears to depend on the specific issues being addressed. Individual investigator grants may be better in particular instances: when the research topic is narrowly defined, for core research, and for exploratory research. Research-center grants are effective for addressing complex, interdisciplinary problems.

The committee concludes that the program is reasonably balanced and that it is important for it to continue to maintain such a balance to address the agency's overall mission.

Does the program award grants expeditiously?

Through presentations by EPA staff and interviews with STAR project officers, the committee obtained substantial information about the process that the STAR program uses in awarding grants. That information is summarized in Chapter 2.

It takes 1-2 years from the initial announcement of the RFA to the time when grants are awarded (see Table 2-7). To someone inexperienced with federal government procedural requirements, the grant-awards process might appear unduly bureaucratic and protracted. However, there appear to be few opportunities for reducing this processing time substantially without compromising the program's efforts. For instance, processing time ensures that the RFAs are widely disseminated and that the award process avoids any appearance of conflict of interest. However, the committee learned that NCER is hiring additional staff to assist in the grant awards process to shorten processing time.

Evaluating the efficiency of the process would be much easier if the program established performance benchmarks that indicated the duration of the various steps in an expeditious grants-award process and kept track of the extent to which the program met the benchmarks. In addition to the time required to complete the various steps in awarding a grant (see Table 2-7), the benchmarks could include the number of progress-review meetings grantees participate in, how long it takes to publish a project's annual report on the STAR Web site, and the number of projects being completed according to the original schedule. The benchmarks should be established by using information obtained from other research-sponsoring organizations and EPA's own experience in managing the STAR program. Such benchmarking provides more useful information than process times alone, because it incorporates a comparison of process times with reasonable expectations.

The committee concludes that NCER has established a reasonably expeditious process for awarding grants but recommends that the STAR program establish benchmarks that measure a successful grants-management process and that it keep track of the percentage of the program's grants that meet these benchmarks.

Does the program have a process to demonstrate the communication of individual grant results in the professional literature?

The STAR program collects information on the publications of its investigators and makes this information readily available and searchable on its Web site. However, the information appears to be collected primarily during the period when the investigators are being supported by STAR grants and not necessarily after grants have been completed.

The committee commends the STAR program for making the bibliometric information it collects readily available to the public. The committee encourages STAR to continue to gather and update bibliometric information from its investigators, even after the research has been completed.

Is there a process in place for reviewing the performance of individual investigators and research centers?

The STAR program does not have a formal process in place for assessing the performance of its principal investigators and research centers. The program's project officers are primarily responsible for monitoring individual projects and for ensuring that they comply with grant requirements, but not for evaluating the content of the research.

However, the program has established mechanisms for peer review of research performance. For individual investigators these mechanisms include requiring that researchers submit annual progress reports that are published on the NCER Web site and participate in progress review meetings. Additional efforts to monitor the performance of individual grantees might not be worth while. Not only would they impose additional costs on the agency and the researchers, but the grant would probably be largely completed by the time any problems were identified. The main concern of the program would appear to be to avoid awarding additional grants to investigators who had a history of performing badly. Interviews with project officers indicated that the program has no mechanism for collecting such information.

In addition to the requirements regarding annual reports and progress review meetings, the grants to research centers also require that the centers establish external advisory committees, comprising academics and government officials, who meet annually to provide guidance and evaluate research progress. In addition, EPA's Scientific Advisory Board did conduct a review of the PM centers and could undertake other similar reviews (EPA/SAB 2002; see Appendix B).

The committee concludes that the processes EPA has established for providing peer review of the performance of individual investigators and research centers is adequate. Such peer review procedures may not ensure performance, but it is highly unlikely that the costs and disruptions that additional review efforts would impose on the agency and the researchers would be justified by any resulting performance improvements.

Product Metrics

Is the program funding relevant research that otherwise would not be funded?

On the basis of interviews with agency staff and researchers who have received STAR support, the committee concludes that much of the mission-related research conducted through individual investigator and center awards would not have been possible without the STAR program, inasmuch as these research projects are not funded by other agencies. For instance, EPA is one of the few agencies that provide extramural funding for examining the effects of endocrine disruptors on wildlife (E. Francis, EPA, Washington, D.C., personal commun., August 6, 2002). In addition, STAR's ecologic-indicators program is the primary source of support for research on the development of water-quality indicators for biologic monitoring (B. Levinson, EPA, Washington, D.C., personal commun., August 4, 2002).

Does the program have a schedule for the products it intends to produce, and how well does it adhere to that schedule?

Although no such schedule was presented to the committee, OMB has requested that agencies prepare such a schedule in its guidelines for evaluating research programs under GPRA (OSTP/OMB 2002), and such specification of program outputs is a normal component of budget preparations. Those documents, however, are not usually made available to the public.

Because of the STAR program's heavy reliance on its Web site for communicating its activities to the public, placing such a schedule on its Web site might substantially enhance the efficiency of the program's communication efforts. The committee, on the basis of discussions with EPA officials, learned that EPA program offices may not always be as aware of recent postings on NCER's Web site as they should be. If STAR listed its expected products on NCER's Web site and allowed users of the site to indicate that they would like to be notified when an expected product became available, the likelihood of being informed about issues could be substantially increased.

The committee recommends that the program post a schedule of expected products on its Web site and allow members of the public to indicate whether they would like to be notified when products become available. This not only would provide the public with an opportunity to observe the performance of the program but also would improve the efficiency of its communication efforts.

To what extent are site-specific studies designed to be replicated at other locations?

Some of the STAR research efforts are focused on developing analytic tools or measures of environmental health for particular geographic areas. A major purpose of such grants is to develop approaches that could be adopted in other locations.

The committee was presented with information about the extent to which biologic sampling protocols were replicated at 822 reference sites in

13 western states—work supported by the U.S. Forest Service and EPA (Gilman 2002). The STAR program also supported a pilot program in the mid-Atlantic states to "develop methods to transfer STAR grant results to environmental decision-makers, test these methods in the mid-Atlantic region, and evaluate the feasibility of using the methods in other regions" (Bradley 2002). At the time the committee was briefed on this effort, limited results were available, although the program had identified a number of regional and local decision-makers who "have found the results of the STAR program valuable" (Bradley 2002).

The pertinent questions that reviews of such projects should be addressing are the extent to which the research projects were intended to be replicable, what efforts have been made to promote such replications, the extent to which replications have occurred, and what steps could be taken to promote their occurrence. The committee recommends that the program continue its efforts to collect information to answer those questions.

FELLOWSHIP PROGRAM

The STAR fellowship program is a small component of the overall STAR program whose goals and objectives differ from those of the main research grants program. The fellowship program is an important contribution to the nation's effort to train and "encourage promising students to obtain advanced degrees and pursue careers in environmentally related fields" (EPA 2002b) and to develop the next generation of environmental scientists. The program is the only federal fellowship program designed exclusively for students pursuing advanced degrees in environmental sciences and engineering. It is highly competitive: only 10% of applicants receive funding. Because the fellowship program has been important in encouraging and maintaining a strong interest in environmental science and engineering, the committee considers that the program should be continued and funded.

Until 2002, the STAR program was funding 100-125 fellows per year. However, the president's FY 2003 budget did not contain funding for the program, so no new fellowships were awarded in 2002. (NOTE: The FY 2003 Omnibus Appropriations Act, signed February 2003, appropriated $9.75 million for the STAR fellowship.)

Although the program publishes on the NCER Web site information about all the students receiving fellowships, it does not gather systematic information to track the status of past and currently funded fellows to assess the impact of the STAR program on their careers. To gather information on the influence of the fellowship program, the committee contacted more than 100 STAR fellows who were initially funded in 1995 and 1996 and who would have completed their graduate work. Of the fellows contacted, over 95% indicated high satisfaction with the program. Additional information gathered by the committee permitted it to evaluate the following metrics.

Does STAR have a process for ensuring the selection of high-quality fellows?

The committee gathered information on how the STAR program selects fellows through discussions with people within EPA's peer-review division.

The STAR program's process for selecting high-quality fellows ensures the competitiveness of the program. The program publishes the announcement for the fellowships widely, including posting it on the NCER Web site and distributing it via e-mail to obtain a large pool of applicants. Prospective fellows submit applications that are evaluated in an independent peer-review process that is similar to the one for reviewing individual investigator and center grants (see Chapter 2 for details). Fellowship applications are evaluated according to criteria that includes their academic and employment records, course of proposed research, and potential for success. To receive continued funding, STAR fellowship recipients are required to remain in good academic standing. A fellowship may be terminated if the EPA project officer determines that the fellow is not performing up to the standards of the program.

On the basis of the wide dissemination of the fellowship applications, the peer-review process used for selecting applicants, and the low percentage of applicants who receive support, the committee concludes that the STAR fellowship program ensures a process for selecting high-quality applicants.

What percentage of fellowship recipients obtain their advanced degrees?

Although the program apparently does not collect the information required to assess this metric, the committee's contacts with more than 100 STAR fellows who received funding in the early years of the program indicated that nearly all have already completed their research and graduated in their degree programs. That is a testament to the quality of the process for selecting fellows and to the success of the program. The committee suggests that if the fellowship program continues, EPA may want to collect metrics on the fellows, including the number graduating and positions held, to document the success of the program.

How many fellowship recipients who have completed their graduate work are working in environmental science?

The program has not collected the information required to assess this metric, but nearly 90% of the fellows contacted by committee members were employed in the environmental science field. About 10% of those were working in government, over 55% chose to remain in academe, and the remainder were working in industry, in consulting, or in nonprofit organizations. Many stated that the fellowship was extraordinarily valuable in assisting them to advance their work in environmental science (see Box 5-4).

The committee suggests that if the fellowship program continues, EPA may want to collect information on the careers pursued by fellowship recipients to document the success of the program.

How many fellowship recipients have completed their degree programs with at least one peer-reviewed publication?

Nearly 80% of the STAR fellows contacted by committee members indicated that they had at least one peer-reviewed publication as a result of their research funded through the fellowship program. That is an indication of the quality of the fellows and the quality of their research. The committee suggests that if the fellowship program continues, EPA may want to collect information on publications and other products of fellowship recipients to document the success of the program.

CONCLUSIONS

- The committee conducted an evaluation of the quality, relevance, and performance of the STAR program, as set forth in the recent OMB research and development criteria, using metrics that grew out of its review of information available from EPA and of metrics used by EPA and other organizations. The metrics, which are both quantitative and qualitative, assisted the committee in forming judgments regarding the scientific merit of the program and its impact on the agency.
- The committee was able to evaluate the program's process better than its results. Evaluation of research results requires a substantial lapse in time, in that it takes 3-5 years, or more, from the initiation of laboratory or field experiments until the analysis and publication of research results. Considerably more time must elapse to view the impact of the published research on the scientific and regulatory community. Advances in research

BOX 5-4 Responses from STAR Graduate-Student Fellows Regarding the Value of the STAR Fellowship

"Yes, it was valuable in the sense that it helped my mentor pay me as a graduate student. It was valuable to me as an initial award on my CV that I have followed up with additional fellowships throughout the years. This gives me a track record of receiving funding. This track record has and will continue to assist me in my future endeavors to obtain funding/jobs. I believe there are not that many opportunities for graduate students to obtain independent funding and the STAR program was valuable in that sense."

"Yes, The STAR fellowship gave me the freedom to expand my thesis topic into very interesting directions that led to several publications in diverse journals."

"It was an invaluable resource. It freed me from teaching so that I could focus on my Ph.D. research. I believe I was able to finish sooner and investigate more than I would have otherwise. It enabled me to buy textbooks to learn about the field that I was working in. It enabled me to attend conferences that I would not have otherwise attended. By attending these conferences I met others in the field, including other graduate students I am still in contact with, and I gained confidence in myself and my abilities. After graduation, I received a National Science Foundation postdoc and went to Rensselaer Polytechnic Institute (RPI) for three years. Without the EPA fellowship, I might not have had the confidence or the ability to apply for the postdoc and write a successful proposal."

"Yes, very much so. I am presently working on the same system answering additional questions raised during my Ph.D. Not only did the fellowship allow me to complete my dissertation, it has in part spawned an entire research program in the evolution of development of stickleback here at the university. We presently have approximately 10-12 people working on this project, and are collaborating with labs at SUNY Stony Brook, Stanford, and Clark University."

"It was very valuable; in particular having a research budget gave me a lot more independence and flexibility in designing and carrying out my dissertation research. The program is also well known and respected. When I interviewed for jobs after graduation, I think my having received the STAR fellowship strengthened my position considerably."

tend not to be accomplished by the completion of individual research grants, but rather through the combined impacts of multiple research projects on a specific field.

- The committee recognizes that there have already been a substantial—some might say excessive—number of reviews of the STAR program. Most have focused on the administration and operation of the program. The committee is concerned that reviewing STAR too frequently has the potential to be damaging, in that it may divert necessary financial and personnel resources away from the program. The committee therefore recommends that STAR and ORD consider an evaluation structure for conducting future reviews by independent panels of experts, comprising individuals with the appropriate scientific, management, and policy backgrounds. Expert reviews are the best method of evaluating the quality of a research program, and having a structured framework would probably reduce the number of ad hoc, unplanned, and uncoordinated reviews.
- The STAR program funds important research that is not conducted or funded by other agencies. The STAR program has also made commendable efforts to leverage funds through establishment of research partnerships with other agencies.
- Although it is still too early for comprehensive evaluations of the research results of the STAR program, some STAR research efforts have already substantially improved the scientific foundation for decision making and the results produced by STAR investigators have been widely published in peer-reviewed journals.
- It is appropriate for EPA's research efforts to incorporate a balance between core and problem-driven research and a balance between ecologic and health-effects research. STAR research improves the knowledge base required to make sound environmental decisions, and this includes both core and problem-driven research. A balance between human-health and ecologic research is necessary, particularly because much of what is termed ecologic research actually elucidates the processes by which environmental stressors affect both humans and other ecosystems.
- The committee encourages the STAR program to continue funding research that explores future environmental problems within its overall research portfolio. Research devoted to potential environmental threats may help to avoid or reduce the impact of such threats or at the very least put into place the scientific capacity to address them.
- Although the STAR program has used several methods to report on the results of individual grants and centers, it has yet to produce documents that summarize the "state of the science" or provide a synthesis of research results and describe how the results of a group of grants have moved scientific understanding forward. The production of such reports, using outside experts where appropriate, can be extremely useful for targeting gaps in

knowledge and communicating the state of the science to the program's diverse users and audiences. The committee considers the increased production of such reports to be an important improvement that should be made in the STAR program. The appropriate type of state of the science or research synthesis document will depend on the intended audience. Because the STAR supported research often complements that being done elsewhere in ORD, and sometimes in other agencies, the integration and synthesis of research results is a larger issue that in many cases cannot be taken solely by the STAR program and must be addressed by ORD or EPA.

- The STAR program has been commendably aggressive in experimenting with innovative approaches to communicating the results of its funded research to a wide variety of users and audiences, but its success in these efforts has been uneven. EPA and the STAR program have various mechanisms for communicating with the STAR user community. However, the STAR program has not developed an effective strategy for communicating to a wider user community, including state, tribal, local, and international environmental agencies and the public; most of the emphasis has been on the scientific community and the program offices. In some cases, the effective dissemination of results should be primarily STAR's responsibility. In other cases, STAR's contributions will be only one component of a larger research effort, and the primary dissemination responsibility should lie within ORD or EPA.
- The fellowship program is an important and valuable component of the STAR program for EPA and the nation. It ensures a continuing supply of graduate students in environmental science and engineering who provide a strong foundation for the nation's environmental research and management efforts. The program has been important in encouraging and maintaining strong interest in environmental science and engineering.

RECOMMENDATIONS

- The committee recommends that NCER institute a structured system of program-level reviews as its primary mechanism for evaluating the STAR program. The improved information-collection efforts (discussed in Chapter 5) should be used to support such reviews.
- The committee recommends that STAR and ORD continue to work to produce state-of-the-science and research-synthesis documents. These are important for identifying critical information gaps and communicating the state of knowledge on a particular issue to the many users and audiences interested in this information.

- The committee commends EPA for its efforts to communicate with its diverse users and audiences and recommends that STAR and EPA continue and, where appropriate, expand such outreach efforts. The likely audiences of research results should be identified early in the research planning process, explicitly identified in RFAs, and considered throughout the research implementation process; and a coherent strategy should be developed for disseminating research results when they become available.
- STAR program funding should be maintained at 15-20% of the overall ORD budget, even in budget-constrained times. However, budget planners should clearly recognize the constraints of not having inflation escalators to maintain the level of effort of the entire program.
- EPA should continue its efforts to attract "the best and the brightest" researchers to compete for STAR funding.
- Given the nation's continuing need for highly qualified scientists and engineers in environmental research and management, the STAR fellowship program should be continued and funded.

REFERENCES

Blanck, H.M., M. Marcus, P.E. Tolbert, C. Rubin, A.K. Henderson, V.S. Hertzberg, R.H. Zhang, and L. Cameron. 2000. Age at menarche and tanner stage in girls exposed in utero and postnatally to polybrominated biphenyl. Epidemiology 11(6):641-647.

Bradley, P. 2002. Pilot Program Overview. Presentation at the First Meeting on the Review of EPA's Research Grants Program, March 19, 2002, Washington, DC.

Bryan, E. 2002. Peer Review Process for EPA's National Center for Environmental Research's Science to Achieve Results Program. Presentation at the Third Meeting on the Review of EPA's Research Grants Program, June 6, 2002, Washington, DC.

Butler, T.J., G.E. Likens, and B.J.B. Stunder. 2001. Regional-scale impacts of Phase I of the Clean Air Act Amendments in the USA: The relation between emissions and concentrations, both wet and dry. Atmos. Environ. 35(6):1015-1028.

Cass, G.R., L.A. Hughes, P. Bhave, M.J. Kleeman, J.O. Allen, and L.G. Salmon. 2000. The chemical composition of atmospheric ultrafine particles. Philosophical Transactions: Mathematical, Physical and Engineering Sciences 358 (1775):2581-2592.

Clarke, R.W., P.J. Catalano, P. Koutrakis, G.G. Murthy, C. Sioutas, J. Paulauskis, B. Coull, S. Ferguson, and J.J. Godleski. 1999. Urban air particulate inhalation alters pulmonary function and induce pulmonary inflammation in a rodent model of chronic bronchitis. Inhal. Toxicol. 11(8):637-656.

Cooke, J.B., and D.E. Hinton. 1999. Promotion by 17beta-estradiol and beta-hexachlorocyclohexane of hepatocellular tumors in medaka, Oryzias latipes. Aquat. Toxicol. 45(2):127-145.

Deegan, L.A., J.T. Finn, and J. Buonaccorsi. 1997. Development and validation of an estuarine biotic integrity index. Estuaries 20(3):601-617.

EPA (U.S. Environmental Protection Agency). 1998a. Ecological Research Strategy. EPA/600/R-98/086. Office of Research and Development, U.S. Environmental Protection Agency, Washington, DC [Online]. Available: http://www.epa.gov/ordntrnt/ ORD/WebPubs/final/eco.pdf [accessed Jan. 13, 2003].

EPA (U.S. Environmental Protection Agency). 1998b. Research Plan for Endocrine Disruptors. EPA/600/R-98/087. Office of Research and Development, U.S. Environmental Protection Agency, Washington, DC [Online]. Available: http://www.epa.gov/ORD/WebPubs/final/revendocrine.pdf [accessed Feb. 20, 2003].

EPA (U.S. Environmental Protection Agency). 1999. Airborne Particulate Matter Research Strategy. EPA/600/R-99/045. Office of Research and Development, U.S. Environmental Protection Agency, Washington, DC. [Online]. Available: http://www.epa.gov/ORD/resplans/Pmstrat7.pdf [accessed Feb. 20, 2003].

EPA (U.S. Environmental Protection Agency). 2000. Evaluation Report: A Decision Making and Valuation for Environmental Policy Interim Assessment. National Center for Environmental Research, Office of Research and Development, U.S. Environmental Protection Agency [Online]. Available: http://es.epa.gov/ncer/science/economics/reviews.html [accessed Jan. 13, 2003].

EPA (U.S. Environmental Protection Agency). 2002a. Report to ORD's Board of Scientific Counselors, Self-Study Update. Prepared by the National Center for Environmental Research, Office of Research and Development, U.S. Environmental Protection Agency. January 2002.

EPA (U.S. Environmental Protection Agency). 2002b. Fall 2002 Science to Achieve Results Fellowships for Graduate Environmental Study. 2002 RFA. National Center for Environmental Research, Officer of Research and Development, U.S. Environmental Protection Agency.

EPA/BOSC (U.S. Environmental Protection Agency Board of Scientific Counselors). 1998. Program Review of the National Center for Environmental Research and Quality Assurance (NCERQA). Final Report of the Ad Hoc Subcommittee on the Review of NCERQA. Board of Scientific Counselors, Office of Research and Development, U.S. Environmental Protection Agency, Washington, DC. April 30, 1998.

EPA/NSF (U.S. Environmental Protection Agency and National Science Foundation). 2000. Interim Assessment for the Decision Making and Valuation for Environmental Policy Grants Program. Final Report. Prepared for National Science Foundation and U.S. Environmental Protection Agency, by Aspen Systems Corporation. April 17, 2000.

EPA/SAB (U.S. Environmental Protection Agency Science Advisory Board). 2001. The Science to Achieve Results (STAR) Water and Watersheds Grants

Program: An EPA Science Advisory Board Review. A Review by the Ecological Processes and Effects Committee (EPEC) of the EPA Science Advisory Board. EPA-SAB-EPEC-02-001. Science Advisory Board, U.S. Environmental Protection Agency, Washington, DC [Online]. Available: http://www.epa.gov/science1/fiscal02.htm [accessed Jan. 13, 2003].

EPA/SAB (U.S. Environmental Protection Agency Science Advisory Board). 2002. Interim Review of the Particulate Matter (PM) Research Centers of the USEPA: An EPA Science Advisory Report. A Review by the PM Research Centers Interim Review Panel of the Executive Committee of the U.S. EPA Science Advisory Board (SAB). EPA-SAB-EC-02-008. Science Advisory Board, U.S. Environmental Protection Agency, Washington, DC. May 2002 [Online]. Available: http://www.epa.gov/science1/fiscal02.htm [accessed Jan. 13, 2003].

EPA/SAB/BOSC (U.S. Environmental Protection Agency Advisory Board and Board of Scientific Counselors). 2000. A joint SAB/BOSC report: Review of the Science to Achieve Results (STAR) program. EPA-SAB-EC-00-008. Science Advisory Board, Board of Scientific Counselors, U.S. Environmental Protection Agency [Online]. Available: http://www.epa.gov/sab/pdf/ec0008.pdf [accessed Jan. 13, 2003].

GAO (U.S. General Accounting Office). 2000. Environmental Research: STAR Grants Focus on Agency Priorities, But Management Enhancements Are Possible: Report to the Chairman, Subcommittee on VA, HUD, and Independent Agencies, Committee on Appropriations, House of Representatives. GAO/RCED-00-170/B-142370. U.S. General Accounting Office, Washington, DC.

Gilman, P. 2002. Presentation at the First Meeting on the Review of EPA's Research Grants Program, March 19, 2002, Washington, DC.

Hawkins, M.B., J.W. Thornton, D. Crews, J.K. Skipper, A. Dotte, and P. Thomas. 2000. Identification of a third distinct estrogen receptor and reclassification of estrogen receptors in teleosts. PNAS 97(20):10751-10756.

Hughes, C., W. Foster, S. Chan, L. Platt, S. Thompson, S. Hubbard, A. DuBose, and L. Tyrey. 2001. Extrapolation of rodent studies on amniotic fluid contaminants to human populations. Human and Ecological Risk Assessment 7(5):979-1002.

Jenkins, R., R.A. Angus, H. McNatt, W.M. Howell, J.A. Kemppainen, M. Kirk, and E.M. Wilson. 2001. Identification of androstenedione in a river containing paper mill effluent. Environ. Toxicol. Chem. 20(6):1325-1331.

Juliano, J., and M.D. Sobsey. 1997. Simultaneous concentration of Cryptosporidium, bacteria and viruses from water by hollow fiber ultrafiltration. In: Proceedings of the 1997 Water Quality Technology Conference, American Water Works Association, Denver, CO, 1997.

Locci, A.B., and J.F. Koonce. 1999. A theoretical analysis of food web constraints on walleye dynamics in Lake Erie. Pp. 497-510 in State of Lake Erie: Past, and Future, M. Munawar, T. Edsall, and I.F. Munawar, eds. Leiden, The Netherlands: Backhuys Publishers.

Lovett, E, R.W. Clarke, R.L. Verrier, P. Koutrakis, J. Lawrence, J.M. Antonini, and J.J. Godleski. 1999. Rat cardiovascular dysfunction prior to death during exposure to concentrated ambient air particles. The Toxicologist 48(1-S):297.

Michalek, J.L., and J.L. Colwell. 2000. Monitoring development in Southwest Florida (1973-1995) using Landsat data and a hybrid change detection technique. Natural Areas Journal April 2000.

NRC (National Research Council). 1998. Research Priorities for Airborne Particulate Matter. 1. Immediate Priorities and a Long-Range Research Portfolio. Washington, DC: National Academy Press.

NRC (National Research Council). 1999a. Evaluating Federal Research Programs: Research and the Government Performance and Results Act. Washington, DC: National Academy Press.

NRC (National Research Council). 1999b. Research Priorities for Airborne Particulate Matter. 2. Evaluating Research Progress and Updating the Portfolio. Washington, DC: National Academy Press.

NRC (National Research Council). 2001. Research Priorities for Airborne Particulate Matter. 3. Early Research Progress. Washington, DC: National Academy Press

OSTP/OMB (Office of Science Technology and Policy/Office of Management and Budget). 2002. FY 2004 Interagency Research and Development Priorities. Memorandum for the Heads of Executive Departments and Agencies, from John Marburger, Director, Office of Science and Technology Policy, and Mitchell Daniels, Director, Office of Management and Budget, The White House, Washington, DC. May 30, 2002.

Pope, C.A., R.T. Burnett, M.J. Thun, E.E. Calle, D. Krewski, K. Ito, and G.D. Thurston. 2002. Lung cancer, cardiopulmonary mortality, and long-term exposure to fine particulate air pollution. JAMA 287(9):1132-1141.

Preuss, P.W. 2002. National Center for Environmental Research, History, Goals, and Operation of the STAR Program. Presentation at The First Meeting on Review EPA's Research Grant Program, March 18, 2002, Washington, DC.

Voinov, A., R. Costanza, L. Wainger, R. Boumans, F. Villa, T. Maxwell, and H. Voinov. 1999. Patuxent landscape model: Integrated ecological economic modeling of a watershed. Environmental Modelling and Software 14(5):473-491.

Xing, L., W.J. Welsh, W. Tong, R. Perkins, and D.M. Sheehan. 1999. Comparison of estrogen receptor alpha and beta subtypes based on comparative molecular field analysis (CoMFA). SAR QSAR Environ. Res. 10(2-3):215-237.

Zanobetti, A., and J. Schwartz. 2001. Are diabetics more susceptible to the health effects of airborne particles? Am. J. Respir. Crit. Care Med. 164(5):831-833.

Appendixes

Appendix A

Biographic Information on the Committee to Review EPA's Research Grants Program

Harold Mooney (Chair) is the Paul S. Achilles Professor of Environmental Biology at Stanford University. He earned his Ph.D. from Duke University. Dr. Mooney's research interests include physiologic plant ecology and ecosystem sciences, the study of adaptations of plants to diverse environments, atmosphere-vegetation interaction, ecosystem functioning of biodiversity, and invasion biology. He has received numerous awards and honors, including membership in the National Academy of Sciences and the American Academy of Arts and Sciences and fellowship in the American Association for the Advancement of Science. Dr. Mooney is secretary general of the International Council for Science and cochair of the Science Panel for the Millennium Ecosystem Assessment. He has extensive service with the National Research Council, including membership on the Committee on Ecosystem Management for Sustainable Marine Fisheries and on the Board on Environmental Studies and Toxicology.

Raymond Loehr (Vice Chair) is the H.M. Alharthy Centennial Chair and a professor of civil engineering at the University of Texas in Austin. He received a Ph.D. in sanitary engineering from the University of Wisconsin. Dr. Loehr's research interests include environmental health engineering, water and wastewater treatment, hazardous-waste treatment, industrial-waste management, and land treatment of wastes. He is a member of the National Academy of Engineering and has served in several National Research Council activities, including the Committee on Remediation of PCB-Contaminated Sediments, the Committee on Research and Peer Review in EPA, and the Board on Environmental Studies and Toxicology. Dr. Loehr was chair of the EPA Science Advisory Board from 1988 to 1994; he currently serves on the board's Executive Committee and chairs its Research Strategies Advisory Committee.

Anders Andren is director of the Sea Grant Institute of the University of Wisconsin-Madison and director of the University of Wisconsin Water Resources Institute. He is also professor of environmental chemistry and technology in the Department of Civil and Environmental Engineering. Dr. Andren received his Ph.D. in chemical oceanography from the Florida State University. His research interests include aquatic and atmospheric chemistry, geochemistry, and analytical chemistry. Dr. Andren is a member of the National Oceanic and Atmospheric Administration Advisory Council of Senior Research Managers and chair of the Sea Grant Program Mission Committee. From 1994 to 2000, he served as a member of the National Research Council Committee on Research and Peer Review in EPA.

Edwin H. Clark, II, is president of Clean Sites Inc. in Alexandria, Virginia. He is the former secretary of natural resources and environmental control for the state of Delaware, vice president of the Conservation Foundation, and associate assistant administrator for pesticides and toxic substances in the Environmental Protection Agency. He holds a Ph.D. in applied economics from Princeton University. He has served as a member of the National Research Council Board on Environmental Studies and Toxicology and on several committees, including the Committee on Risk-Based Criteria for Non-RCRA Hazardous Waste.

Costel Denson is professor in the Department of Chemical Engineering at the University of Delaware. He received his Ph.D. from the University of Utah. His research has focused on the rheology and processing of polymeric materials. Dr. Denson is a member of the National Research Council's Committee on Air Quality Management in the United States and has served on the Ford Foundation Minority Predoctoral Review Panel on Engineering. He has also served as the chair of the Board of Scientific Counselors for the Environmental Protection Agency and is a member of the National Science Foundation Advisory Committee for Environmental Research and Education.

John Elston is retired from serving as administrator for the Office of Air Quality Management in the New Jersey state Department of Environmental Protection. Mr. Elston holds an MS in environmental science from Rutgers University. As administrator he directed the planning, preparation, and tracking of progress for the state implementation plan for attainment and maintenance of the national ambient air quality standards under the Clean Air Act and plans for air-pollutant monitoring systems for the state.

Carol Henry is the vice president for science and research at the American Chemistry Council (ACC, formerly the Chemical Manufacturers Association). She directs and manages ACC's Long-Range Research Initiative that is designed to study the potential impacts of chemicals on health and the environment. Dr. Henry earned her Ph.D. in microbiology from the University of Pittsburgh. Her interests include the scientific foundation for assessing risks to health and the environment and the management of scientific research programs. Dr. Henry has served in senior management positions at the American Petroleum Institute, the Department of Energy, the California Environmental Protection Agency, and the International Life Sciences Institute's Risk Science Institute. She serves as a member of the National Research Council's Board on Environmental Studies and Toxicology; as a member of the Science Advisory Board of the Strategic Environmental Research and Development Program of the U.S. Department of Defense, Department of Energy, and Environmental Protection Agency (EPA); and as a consultant to EPA's Science Advisory Board.

Martha A. Krebs is president of Science Strategies, a consulting firm. She served most recently as associate vice chancellor for research at the University of California, Los Angeles (UCLA) and as the founding director of the California NanoSystems Institute at UCLA and UC, Santa Barbara. Dr. Krebs received her Ph.D. from the Catholic University of America. From 1993 to 2000, she served as assistant secretary of energy and director of the Department of Energy (DOE) Office of Science, where she was responsible for the $3 billion basic-research program that underlay the department's energy, environmental, and national-security missions. Before that, Dr. Krebs was associate director for planning and development at DOE's Lawrence Berkeley National Laboratory. She serves on the Board of Trustees for the Institute for Defense Analyses and is a member of the National Research Council's Board on Energy and Environmental Systems. Dr. Krebs is a member of the American Physical Society, a Fellow of the American Association for the Advancement of Science, and a Fellow of the Association of Women in Science.

Richard Lee is professor of oceanography at the Skidaway Institute of Oceanography. He received his Ph.D. in marine biology from the Scripps Institute of Oceanography. His research interests include bioremediation, DNA damage and embryo developments in grass shrimp exposed to contaminants, and blue crab disease in coastal Georgia. Dr. Lee served on the National Research Council's Committee on Marine Salvage.

Gerald van Belle holds joint appointments as professor in the Department of Biostatistics and in the Department of Environmental Health at the University of Washington. He chaired the Department of Environmental Health from 1991 to 1998. Dr. van Belle received his Ph.D. in mathematical statistics from the University of Toronto. His research interests have focused on the use of statistics to study various environmental health issues related to Alzheimer's disease, exposure to air pollutants, and drinking-water quality. From 1993 to 1996, Dr. van Belle served as a member of the National Research Council's Board on Environmental Studies and Toxicology. In 1999, he was a member of the Particulate Matter Center Review Panel for the Environmental Protection Agency.

Terry Young is senior consulting scientist at Environmental Defense in Oakland, California. She received her Ph.D. in agricultural and environmental chemistry from the University of California, Berkeley. Dr. Young is a member of the Environmental Protection Agency's Science Advisory Board (SAB) and serves as chair of the Ecological Processes and Effects Committee of the SAB. At Environmental Defense, she manages projects on water and sediment quality, wetland and riverine habitats, and the development of economic incentives for pollution control.

Lauren Zeise is chief of the reproductive and cancer hazard assessment section of the California Environmental Protection Agency. She received her Ph.D. from Harvard University. Dr. Zeise's research focuses on modeling human interindividual variability in metabolism and risk. She has served on advisory boards of the Environmental Protection Agency, the World Health Organization, the Office of Technology Assessment, and the National Institute of Environmental Health Sciences. She has also served on several National Research Council committees, including the Committee on Risk Characterization, the Committee on Comparative Toxicology of Naturally Occurring Carcinogens, and the Committee on Copper in Drinking Water. Dr. Zeise is a member of the Committee on Toxicology and the Board on Environmental Studies and Toxicology.

Appendix B

Previous Reviews of the STAR Program

This appendix summarizes the conclusions and recommendations of the various external reviews that have been conducted on the Environmental Protection Agency (EPA) Science To Achieve Results (STAR) program and the agency's official responses to the reviews.

1997 REVIEW BY ORD BOARD OF SCIENTIFIC COUNSELORS

The first review of the STAR program was conducted in 1997 by EPA's Office of Research and Development (ORD) Board of Scientific Counselors (BOSC). BOSC conducted a series of management and programmatic reviews of ORD's three major laboratories and two centers. One of the centers was the National Center for Environmental Research and Quality Assurance (NCERQA), now known as the National Center for Environmental Research (NCER), which had responsibility for overseeing and administering the STAR program. BOSC concluded that NCERQA was an organization of vital importance to ORD, EPA, and the national objective of improving fundamental knowledge for environmental assessment and management. It further concluded that NCERQA had played a key role in refocusing and shaping the new vision for ORD.

In the course of its review of NCERQA, BOSC made several recommendations that touched on the STAR program (EPA/BOSC 1998), including the following:

- "That support and expectations of Project Officers be addressed as a priority issue in NCERQA management planning. More prioritization, and in some cases streamlining of grant and fellowship management, appear necessary.
- That NCERQA continue to expand its cooperation and connections with other federal, private, and international environmental research organizations through joint solicitations, Web site links, and the exchange of ideas and research results.
- That RFA workshop proceedings be expanded to include a record of discussions, exchange of ideas, integration across research projects and their relevancy for environmental decision making.
- That NCERQA continue to support and expand its Web site as a central location for information on the Center, ORD, and other organizations performing related research.
- That NCERQA require investigators to discuss the relevance of their research to EPA as a part of their project summaries."

NCERQA, in a written response (Noonan 1999), addressed all the recommendations and described the actions taken up to that point. The response was judged to be satisfactory.

2000 REVIEW BY A SPECIAL JOINT COMMITTEE ESTABLISHED BY EPA'S SCIENCE ADVISORY BOARD AND ORD'S BOARD OF SCIENTIFIC COUNSELORS

In the latter part of 1999, EPA's Science Advisory Board (SAB) and ORD's BOSC established a joint committee to review the STAR program (EPA/SAB/BOSC 2000). That committee concluded that the STAR program is of vital importance to EPA's mission and to the national objective of improving the knowledge base for environmental assessment and management. The committee concluded further that the STAR program was structured and managed to generate high-quality science by well-qualified scientists on relevant topics as identified in the EPA strategic plan.

The committee's recommendations fell along two general lines: staff resources and information transfer. The committee felt that greater staff resources were required for maximizing the public's return on investment in the STAR program. Coincidentally, that staffing need was identified in BOSC's 1998 review of NCERQA (EPA/BOSC 1998).

Concerning information transfer, the committee felt that greater emphasis and attention needed to be placed on developing and implementing the tools, management processes, and procedures for ensuring that the information from and results of the STAR program are rapidly and effectively transferred to agency users. Ten recommendations were made:

- "The Agency should provide additional information in RFAs on research goals and objectives and on budget and relevancy criteria that will be used to evaluate proposals.
- The Agency should take steps to accelerate the peer review process for STAR results.
- The Agency should select several STAR research grants as case examples and evaluate the effectiveness of the coordination with the relevant client offices and the degree to which the awards are supporting the Agency's strategic goals.
- The Agency should consider means of strengthening communications between Agency program staff and STAR grant recipients.
- The Agency should assess how well the needs and issues of the regional offices are factored into the STAR planning process and consider additional mechanisms for ensuring adequate regional involvement in STAR Program activities.
- The Agency should request feedback on the success of the program review workshops and should expand the workshop proceedings to include a record of discussions regarding the relevancy of STAR results to the Agency's research and regulatory agenda and to environmental decision making.
- The Agency should develop and implement a process for periodically assessing the Agency's portfolio in terms of its use of different funding instruments and the reliance on different R&D performers.
- The Agency should continue and expand its partnerships with other agencies and funding organizations.
- The Agency should seek assistance from program evaluation and decision analysis experts to help ORD develop a monitoring and evaluation system for the STAR Program.
- The Agency should budget sufficient resources to secure the services of a qualified, highly respected, and independent organization to conduct and publish an evaluation of the STAR Program's results, effectiveness and impact."

ORD responded to the SAB and BOSC review in a written response from Norine Noonan (Noonan 2000a) in which she acknowledged that the agency had implemented two of the recommendations, including allocating additional staff resources to the STAR program and hiring an additional staff person to assist with the communication of results. In addition, she noted that work had begun on a pilot "state-of-the-science" report on ecologic indicators.

2000 GENERAL ACCOUNTING OFFICE REPORT

The General Accounting Office (GAO) published a report in September 2000 (GAO 2000) that reviewed the STAR grants program with emphasis on three issues: whether funding amounts awarded for the grants align with EPA's strategic goals, ORD's research priorities, and program-office priorities; the extent to which the completed focused grants have provided research that is being used by EPA's program offices; and how ORD could enhance its management of the program to help to ensure that it meets its objectives.

Broadly speaking, GAO found that STAR grant funding had generally been aligned with EPA's, ORD's, and the program offices' broadly defined priorities. It found, however, that EPA's program officials varied in the extent to which they believed that the grants' results were useful to them. GAO noted further that ORD could enhance its management of the program to help to ensure that it meets its objectives.

GAO made three recommendations to enhance the effectiveness of the STAR program. The administrator of EPA and the assistant administrator of ORD, it said, should take the following actions:

- "Track and monitor the grants to ensure that interim and final research results are delivered on time and are made available as soon as possible for use by the program offices.
- Take the additional steps needed to disseminate and communicate STAR research results to the appropriate program officials better. That would require continuing and expanding the efforts already under way to consult with program officers in determining the most effective communication methods.
- Develop program criteria to evaluate the effectiveness of each type of grant—exploratory grants, focused grants, and fellowships. In addition, the criteria should assist EPA in drawing an overall conclusion on whether the grants satisfy the program's overall objectives."

In a letter dated October 17, 2000, the assistant administrator responded to the GAO report (Noonan 2000b) and stated that steps were being taken to implement the recommendations of the report and the recommendations made in the EPA SAB-BOSC study.

2001 SCIENCE ADVISORY BOARD REVIEW OF THE WATER AND WATERSHEDS PROGRAM

In autumn 2001, the Ecological Processes and Effects Committee (EPEC) of EPA's SAB reviewed the water and watersheds component of STAR (EPA/SAB 2001). The EPEC review was the first to focus on the quality and utility of the research funded by STAR in a particular subject. At the time the review was conducted, the STAR water and watersheds program had provided about $36 million over a 5-year period, and research results of multiple grant cycles were available.

EPEC commended the STAR water and watersheds program for the quality of its research, for refining decision tools (primarily computer models), for producing a crop of young researchers with experience in an important environmental field, and for legitimizing transdisciplinary research in the academic community. The committee recommended that the water and watersheds program be continued with the following midcourse corrections (EPA/SAB 2001):

- Refine the requests for proposals so that the resulting research focuses more sharply on information gaps and policy-relevant research topics.
- Begin to synthesize results from the collective body of research funded by the program and disseminate this information in useful forms to the rest of EPA and its partners in state and local agencies.

In addition, EPEC provided examples of metrics that could be used to judge the success of the program in the future.

2000 JOINT REVIEW BY EPA AND NSF

In 2000, the National Science Foundation (NSF) and EPA convened experts from outside the agencies and users of socioeconomic research, including EPA program and regional staff, to conduct an interim assessment of the Decision Making and Valuation for Environmental Policy grants program (DMVEP) (EPA/NSF 2000; EPA 2000). NSF and EPA, which managed the program jointly, were interested in determining whether it was producing results and communicating them effectively. At the time of the review, the DMVEP program had been in operation for 5 years, and about $2 million had been given to support it each year.

The experts who conducted the review concluded that the DMVEP program fills a critical research niche that is not addressed by other research programs and commended the program for advancing the state of knowledge in an underfunded field and helping to develop a new field of study. The experts recommended the following midcourse corrections:

- "Increase outreach and communication efforts, to improve awareness both of funding opportunities and of research findings;
- Continue to support research on both monetary and non-monetizable ecosystem valuation; and
- Encourage research on group and institutional—as well as individual—valuation and decision making for environmental policy."

Although EPA did not respond formally to those recommendations, it did implement many of the recommendations in the 2001 and 2002 DMVEP solicitations.

2002 SCIENCE ADVISORY BOARD REVIEW OF PARTICULATE-MATTER CENTERS

In February 2002, the PM Research Centers Interim Review Panel of EPA's SAB convened a review of the particulate-matter research centers program. At the time of the review, the agency had been funding five PM centers since 1999 at about $8 million per year; about half the grant period had elapsed. The review was intended to provide the agency with guidance as to whether it should continue the concept of the PM research centers beyond FY 2004 or whether there was a better mechanism of generating the research results to inform EPA's decision making on PM (EPA/SAB 2002).

The PM Research Centers Interim Review Panel concluded that the PM centers program had produced benefits beyond those normally found in individual investigator-initiated grants and that the program merited continuation.

The panel identified several advantages that centers offer over traditional investigator-initiated awards, including enhanced flexibility and adaptability, leading to improved timeliness; ability to conduct higher-risk pilot and validation efforts; study designs enhanced by intracenter multidisciplinary integration; and improved leveraging of research resources.

The panel offered several recommendations for improving the centers' program:

- "Focusing the centers' efforts on the most critical PM needs in the new RFAs.
- Development of an informal, but overarching mechanism for providing scientific advice to the centers' program.
- Enhanced opportunities for cross-fertilization of ideas with EPA's intramural researchers and the larger extramural community.
- Enhanced interaction between the research conducted at the centers and ongoing intensive air quality monitoring efforts.
- Providing mechanisms and resources for inter-center integration" (EPA/SAB 2002).

2002 BOARD OF SCIENTIFIC COUNSELORS EVALUATION

In November 2002, ORD's BOSC issued the results of its program review of NCER, one in a series of programmatic reviews the board was conducting of the ORD laboratories and centers in response to a request made in 2000 by Assistant Administrator for Research and Development Henry Longest III (EPA/BOSC 2002). The review was a follow up to the 1997 review by the same organization and relied in part on a "self-study" report prepared by the NCER staff in response to a series of 19 questions put to the organization by BOSC (EPA 2002). The review was carried out by a specially constituted subcommittee of BOSC.

The subcommittee concluded that "the National Center for Environmental Research (NCER) has a strong and dynamic research program that is well connected to the Office of Research and Development (ORD) and the Environmental Protection Agency (EPA) Program Offices." It emphasized NCER's "strong and creative leadership" and the staff's "enthusiasm and professionalism." The report had the following 16 findings and recommendations (EPA/BOSC 2002):

> "Recommendation 1: NCER should proceed with development of its Strategic Plan as soon as possible. The plan can serve as the cornerstone for measuring the health of NCER and determining its future resource requirements.

Recommendation 2: The communications of NCER decisions and actions surrounding research initiation and prioritization decision processes would be enhanced with the development of written documentation of the processes surrounding RFA prioritization, setting of funding levels for research topics, and initiating programmatic changes in the course of MYPs.

Building on some successful efforts to date, additional effort is needed to refine and focus RFA solicitations to ensure that proposers, reviewers, and EPA staff understand the scope and focus of the research area to be addressed and its importance.

Recommendation 3: NCER should develop a strategy or model to gauge the balance between the number of RFAs issued and available funding.

Recommendation 4: The structured, integrative process being used by NCER to identify research topics and conduct research has considerable merit. However, streamlining measures should be taken to reduce the time demands on staff while preserving the essence of an integrative process.

Recommendation 5: Progress of NCER in partnering with other federal agencies has been excellent, firmly establishing it as a leader in environmental research and significantly raising its visibility. This is certainly an accomplishment for which NCER should be commended.

Recommendation 6: Although NCER's budget for social science research is small, NCER is making progress in integrating social science into other STAR program research initiatives. A broader range of social science research is needed, in addition to the focus on economics.

Recommendation 7: NCER should address the issue of research balance of human and ecological areas more routinely, and clarify the rationale (not just the methodology) for the balance selected. This should address the context of balance within EPA, and more globally, considering EPA's contributions among other federal research programs.

Recommendation 8: NCER has achieved commendable progress in developing effective systems for managing its programs and motivating its staff. The BOSC encourages its continuation through, for example, careful tracking of new hires to ensure their continued enthusiasm.

Recommendation 9: Although NCER's unique qualities and capabilities within EPA are apparent and well recognized by those familiar with EPA, there was concern that other federal research programs and legislators may not appreciate these areas of distinction. It is vital that NCER document exactly how it differs from other, related federal and non-federal programs and why it can accomplish things that these other agencies cannot.

Recommendation 10: NCER's leadership clearly recognizes the need for effective communications (and for the benefits of improved program coordination that result from such communications), as evidenced by the enviable record of publications and information bulletins from the STAR program, numbers of "hits" on its Web site, and ambitions to develop new tools.

The ambition of outreach to each of the important and diverse audiences is noble.

However, it is clear that full realization of the goal remains to be achieved, and working toward such a goal will have a large impact on personnel in terms of the nature and amount of effort NCER has to mount.

NCER can best achieve its goals by: (1) intensifying communications between NCER and its many audiences, (2) initiating those communications earlier in the research planning process, and (3) assuring that NCER's research results reach those who are in a position to apply them to health and/or environmental improvement.

NCER should continue to expand its proactive program of education and outreach to be sure that the results of NCER-funded research take their proper role in guiding EPA's regulatory programs, and that Congress exercises its duties in oversight and support.

Recommendation 11: NCER's lack of performance measures is linked to the absence of a strategic plan and a single agency to provide a benchmark for comparison. NCER should complete its Strategic Plan and develop reference points by gathering benchmarks from an eclectic group of agencies whose activities intersect those of NCER.

Recommendation 12: NCER programs can focus on priority areas of research of interest to EPA while meeting the concerns and interests of entities outside the Agency. We encourage NCER to continue to maintain project planning and management approaches that allow integration of broad stakeholder interests and priorities.

Recommendation 13: The case studies presented in the Self-Study Report indicated that budget limitations were the cause of a research program not meeting the expectations of a Program Office (sponsor). NCER is admonished to be cognizant of others barriers (e.g., a too narrowly defined project that overlooks chronic health impacts) that also can contribute to a sponsor's expectations not being met.

Recommendation 14: NCER has a commendable leadership role in seeking development of metrics for quantitative evaluations of research quality and impact. This is a significant area, and is a challenge of sufficient importance to justify allocation of additional personnel time and research funds.

Recommendation 15: NCER and ORD have provided a number of strategic opportunities and quality tools that can be used to feed NCER research results back into the EPA research planning process, providing a means to impact the establishment of research priorities and integration with EPA's mission. It appears that this process is effective but may not yet be optimized, pending more efficient communication and time-management considerations.

Recommendation 16: NCER is doing a good job with its current resources and recognizing its needs. As the Center shifts the responsibilities of staff, the following are offered as suggestions:

Link resource type and quantity to activities defined by the Strategic Plan; Develop innovative approaches to solve paperwork bottlenecks that currently are personnel intensive (the digital processes at NSF and other agencies are recommended starting places); and Develop performance measures for internal and external communication plans."

EPA had not prepared a response to those recommendations by the beginning of December 2002.

REFERENCES

EPA (U.S. Environmental Protection Agency). 2000. Evaluation Report: A Decision Making and Valuation for Environmental Policy Interim Assessment. National Center for Environmental Research, Office of Research and Development, U.S. Environmental Protection Agency [Online]. Available: http://es.epa.gov/ncer/science/economics/reviews.html [accessed Jan. 13, 2003].

EPA (U.S. Environmental Protection Agency). 2002. Report to ORD's Board of Scientific Counselors, Self-Study Update. Prepared by the National Center for Environmental Research, Office of Research and Development, U.S. Environmental Protection Agency. January 2002.

EPA/BOSC (U.S. Environmental Protection Agency Board of Scientific Counselors). 1998. Program Review of the National Center for Environmental Research and Quality Assurance (NCERQA). Final Report of the Ad Hoc Subcommittee on the Review of NCERQA. Board of Scientific Counselors, Office of Research and Development, U.S. Environmental Protection Agency, Washington, DC. April 30, 1998.

EPA/BOSC (U.S. Environmental Protection Agency Board of Scientific Counselors). 2002. Second Program Review of the National Center for Environmental Research (NCER). Board of Scientific Counselors, Office of Research and Development, U.S. Environmental Protection Agency, Washington, DC. November 19, 2002.

EPA/NSF (U.S. Environmental Protection Agency and National Science Foundation). 2000. Interim Assessment for the Decision Making and Valuation for Environmental Policy Grants Program. Final Report. Prepared for National Science Foundation and U.S. Environmental Protection Agency, by Aspen Systems Corporation. April 17, 2000.

EPA/SAB (U.S. Environmental Protection Agency Science Advisory Board). 2001. The Science to Achieve Results (STAR) Water and Watersheds Grants Program: An EPA Science Advisory Board Review. A Review by the Ecological Processes and Effects Committee (EPEC) of the EPA Science Advisory Board. EPA-SAB-EPEC-02-001. Science Advisory Board, U.S. Environmental Protection Agency, Washington, DC [Online]. Available: http://www.epa.gov/science1/fiscal02.htm. [accessed Jan. 13, 2003].

EPA/SAB (U.S. Environmental Protection Agency Science Advisory Board). 2002. Interim Review of the Particulate Matter (PM) Research Centers of the USEPA: An EPA Science Advisory Report. A Review by the PM Research Centers Interim Review Panel of the Executive Committee of the U.S. EPA Science Advisory Board (SAB). EPA-SAB-EC-02-008. Science Advisory Board, U.S. Environmental Protection Agency, Washington, DC. May 2002 [Online]. Available: http://www.epa.gov/science1/fiscal02.htm [accessed Jan. 13, 2003].

EPA/SAB/BOSC (U.S. Environmental Protection Agency Science Advisory Board and Board of Scientific Counselors). 2000. A Joint SAB/BOSC Report: Review of the Science to Achieve Results (STAR) Program. EPA-SAB-EC-00-008. Science Advisory Board, Board of Scientific Counselors, U.S. Environmental Protection Agency. [Online]. Available: http://www.epa.gov/sab/pdf/ec0008.pdf [accessed Jan. 13, 2003].

GAO (U.S. General Accounting Office). 2000. Environmental Research: STAR Grants Focus on Agency Priorities, But Management Enhancements Are Possible: Report to the Chairman, Subcommittee on VA, HUD, and Independent Agencies, Committee on Appropriations, House of Representatives. GAO/RCED-00-170/B-142370. U.S. General Accounting Office, Washington, DC.

Noonan, N.E. 1999. Letter from N. Noonan, Assistant Administrator, Office of Research and Development, U.S. Environmental Protection Agency, Washington, DC, to C. Denson, Chair of BOSC, University of Delaware, Newark, DE. May 19, 1999.

Noonan, N.E. 2000a. An SAB/BOSC Report: Review of the Science to Achieve Results (STAR) Program (EPA-SAB-EC-00-008). Letter from N. Noonan, Assistant Administrator, Office of Research and Development, U.S. Environmental Protection Agency, Washington, DC, to C. Denson, Chair, Board of Scientific Counselors, University of Delaware, Newark, DE. May 26, 2000.

Noonan, N.E. 2000b. Response to GAO Final Report, Environmental Research: STAR Grants Focus on Agency Priorities, but Management Enhancements are Possible (GAO/RCED-00-170). Letter from N. Noonan, assistant administrator, Office of Research and Development, U.S. Environmental Protection Agency, Washington, DC, to N. Gelb, director, Annual Planning and Budget Division, Office of the Chief Financial Officer, GAO, Washington, DC. October 7, 2000.

Appendix C

Performance Measures Used by Other Agencies and Organizations

Many agencies and organizations involved in managing research programs are attempting to develop performance measures to evaluate the quality of their programs. In an effort to understand how metrics can be used to evaluate programs effectively, the committee reviewed evaluation tools used by the National Institute for Standards and Technology (NIST), the U.S. Air Force, Texas, Maine, Kansas, and two academic programs: The National Science Foundation (NSF) Experimental Program to Stimulate Competitive Research (EPSCOR) and the National Sea Grant Program.

Typically, two primary mechanisms drive the use of performance measures: specific legislative requirements and the desire to benchmark outcomes and impacts of a program (J. Melkers, University of Illinois, Chicago, personal commun., June 8, 2002).

PERFORMANCE MEASURES USED TO EVALUATE FEDERAL AGENCIES

The NIST Advanced Technology Program (ATP) and the U.S. Air Force Scientific Advisory Board (SAB) are two examples of federal-agency research programs that have developed specific metrics or evaluation criteria to gauge the performance of their research programs.

National Institute for Standards and Technology Advanced Technology Program

The NIST ATP has developed a complex program-evaluation tool, the business reporting system (BRS). The BRS, which was implemented in 1994, is used to track companies that have received funding through the ATP. It is an impressive evaluation tool that comprehensively evaluates the business and economic impacts of each research project from start to finish (see Box C-1). Companies are asked to respond to a number of detailed surveys before, during, and after their projects are completed. The surveys include questions regarding the commercial application of proposals, business goals, strategies for commercialization and for protecting intellectual property, dissemination efforts, publication in professional journals and presentations at conferences, participation in user associations, public-relations efforts, R&D status, collaborative efforts, impacts on employment, and attraction of new funding.

U.S. Air Force Research Program

Evaluations of the U.S. Air Force research program are conducted by the U.S. Air Force SAB. The first SAB evaluation was conducted in 1991 (R. Selden, U.S. Air Force SAB, personal commun., Jan. 9, 2003). Programs are evaluated for quality and relevance of research, and each directorate is evaluated every 2 years. Typical metrics used to evaluate research programs are university metrics (publications, patents, and peer review) and a grading system that is used to evaluate the various components of the research programs in each directorate on the basis of 10 criteria (see Box C-2). Scores are normalized across the different directorates (Selden 1998).

PERFORMANCE MEASURES USED TO EVALUATE STATE-LEVEL PROGRAMS

Many states have developed science and technology performance metrics to evaluate whether and how research programs are encouraging economic development. State governments tend to be more interested in whether research programs are encouraging economic development than in the quality or value of the research itself. Texas, Kansas, and Maine have developed a process of using performance measures to evaluate their re-

BOX C-1 NIST ATP Performance Metrics

Companies are asked to submit information regarding their progress and economic contributions in the form of four electronically submitted reports:

Baseline report. Companies submit information regarding commercial application of proposals, business goals, strategies for commercialization and for protecting intellectual property, and dissemination efforts. Companies are asked to rank the importance of publishing in professional journals, presenting papers at conferences, participation in user associations, and public-relation efforts (NIST 2002a).

Anniversary report. Companies list major commercial applications identified in the proposal, business goals, progress toward commercialization, R&D status, collaborative efforts, employment impacts, attraction of new funding, strategies for protecting intellectual property, and dissemination efforts. This section asks companies to evaluate the status of their R&D cycle as a result of ATP funding (NIST 2002b). These reports are completed annually.

Close-out report. Companies discuss commercial applications, business goals, early impacts on revenue and cost, future projections on revenue and costs, R&D status, collaboration impacts, impacts on employment, attraction of new funding, strategies for protecting intellectual property, and dissemination plans (NIST 2002c). Reports are completed after the conclusion of projects.

Post-project summary. Companies provide information on postproject affiliation, funding sources, the impact of ATP funding on product development and process capability, anticipated future market activity, and R&D partnering with other organizations (NIST 2002d). Reports are completed 2, 4, and 6 years after projects are completed.

search and science and technology programs. The performance measures are representative of similar evaluations being conducted by other states.

Texas

The primary research effort in Texas is known as the Advanced Re-evaluation efforts for ARP/ATP are coordinated by the Texas Higher Edu-

BOX C-2 U.S. Air Force Scientific Advisory Board Science and Technology Review Evaluation Criteria

1. Science Foundation
- Work described is based on sufficiently understood phenomena.
- Uses best and most recent available science applicable to the problem.
- All scientific issues, together with the work to address those issues, are identified.
- There is a rigorous approach to stated technical problem.
- Distinction is made between innovative design concepts and innovative science.

2. Strategic Vision
- How does technology fit into evolving military capabilities?
- A clear identifiable path exists which connects technology to military capabilities.
- Leadership of scientific community into new research areas of high leverage for the Air Force?
- Commercial technology growth is forecast and planned for incorporation.

3. Focus of Efforts
- Sufficient resources for a critical mass.
- Accountability exists for technical milestones.
- Scope is defined to maximize output.
- Maximum leverage with other programs exists where appropriate.

4. Research Environment
- Quality and capabilities of facilities and equipment.
- Work atmosphere fosters productive interaction and allows for constructive criticism without fear of retribution.
- External experts are consulted (excluding lab contractors); lab is receptive to external ideas.

5. Approach
- Addresses timely delivery of product.
- Leverages similar research from government/industry.
- Teams with the best from government and/or industry.
- Uses or modifies commercial technology.
- Current effort differs in approach from what was done 5 or 10 years ago.
- Maintains a balanced approach between cost and performance.

(Continued)

BOX C-2 *Continued*

6. Innovation
- Applies novel techniques and cross-disciplinary science.
- New concepts/techniques/devices emerged as a result of the technology.
- Reflects "out-of-the-box" thinking.

7. Output
- Results of technology have effective and timely transition.
- Milestones result in "interim" products (prototypes, increased knowledge).
- Technical quality evidenced by awards from technical societies.
- Customer satisfaction evidenced by customer feedback.
- Understand the metrics of success.

8. People
- Qualifications, reputation, and technical productivity compared with other organizations in the same discipline.
- "Top guns" are involved in research.
- Solid mentoring system in place.
- Programs managers are recognized as experts in their fields.

9. Context
- Understand military capability needs and priorities for technology development.
- Technology fits properly in context with similar research.
- Technology addresses the "tall poles."
- There is an awareness of similar research inside/ outside of DOD.

10. Long-term Relevance
- Technology will have a short-term and/or long-term impact on Air Force capabilities, weapon systems, personnel, and environment.
- Technology addresses unique long-term DOD/USAF weapon system or infrastructure needs (technology push).
- Technology provides meaningful improvement to weapon system sustainability (reliability, maintainability).

Source: USAFSAB/USAFRL 1998.

cation Coordinating Board. Progress reports and final reports are used to track the impact of research projects. Surveys are used to gather information about the progress of research. The survey metrics used to evaluate research projects include number of publications and performance of graduate students (see Box C-3) (ARP/ATP 2002; J. Melkers, University of Illinois, Chicago, IL, personal commun., July 8, 2002).

Kansas

The Kansas Technology Enterprise Corporation is responsible for managing grant programs for applied research and for equipment used in science and technology skill training. Examples of metrics used to evaluate programs are ranking of importance of commonly accepted economic development goals (such as, job creation, encouraging technologic innovation and entrepreneurial spirit, and literature review), and literature reviews (Burress et al. 1992).

Maine

Maine's Research and Development Evaluation Project (headed by the Maine Science and Technology Foundation) was asked by the state legislature to undertake a comprehensive 5-year evaluation that focuses on how the state R&D program has evolved and affected R&D industry and the level of innovation-based economic development in the state. *Evaluation of Maine's Public Investment in Research and Development,* a report produced by the Maine Science and Technology Foundation (MSTF 2001), documents each research program that has been evaluated and the processes and methods used to evaluate each program. Performance measures are varied. For instance, the evaluation of the Maine Biomedical Research Program focused on output and outcome measures. Output measures include a plan showing how the funds would be used and the resulting research and economic benefits, peer-review journal articles demonstrating competitiveness of the institution's research, and the amount of funding from outside sources and its use. Outcome measures include an evaluation of the direct and indirect economic impact of the funded research and an assessment of the contribution of the funded research to scientific advancement and the institution's competitive position (MSTF 2001). The foundation has prepared a survey for research institutions to assist them in collecting data for program evaluation (see Box C-4).

BOX C-3 Texas Higher Education Coordinating Board Research-Project Performance Metrics

The following are the questions being asked in 2002 of all persons who have received ARP/ATP funds. Responses are provided electronically.

A. Provide a short (200 word) description of what you did during this project. This description should be written for a lay rather than a highly technical audience.
B-1. Over the term of this project, how many different people (including the PI's) have been supported by this project? Categories for response include men, women, black, hispanic, native american, foreign national.
B-2. Over the term of the grant, how many additional people have worked on, but not been supported by the project? Same categories as for B-1.
C. Identify those students who have worked on this project and graduated from your institution. Identify where they are now working.
D-1. Over the term of the grant, what additional funding has your institution received, directly or indirectly, as a result of participation in this program?
D-2. Over the term of the grant, what additional funding has your institution requested as a result of participation in this program?
E. Over the term of this grant, how many different publications resulted from this project. Categories are refereed journals, conference proceedings, technical reports, book chapters, and other. For actual publications, full citations are requested.
F. Brief "success" stories are requested.
G. Briefly describe any "industrial" or commercial connections your project has.
H. Briefly describe what you have done to effect technology transfer of the work done in this project.

Subsequent questions are related to information about interaction with actual or potential collaborators, commercialization, knowledge utilization, and possible licensing opportunities.

Source: ARP/ATP 2002.

PERFORMANCE EVALUATION MEASURES IN ACADEME

Program evaluation is also used widely in academe, particularly in programs that involve improving educational opportunities and academic

BOX C-4 Draft Survey for Maine Research Institutions (Revised February 1, 2002)

1. Name of Research Institution:
2. Name of Person Completing Survey:
   Position:
3. Institutional Capacity:
   A. Increase in number of enrolled science and engineering graduate students attributable to state R&D funding
   B. Increase in number of science and engineering graduate degrees awarded attributable to state R&D funding
   C. Number of new degree programs established as a result of state R&D funding
   D. New and/or renovated R&D space available as a result of state R&D funding
   E. Value of new facilities and fixed equipment acquired as a result of state R&D funding
   F. Number of new FTEs hired as a result of state R&D funding
   G. Major (purchase price > $50,000) new research equipment acquired as a result of state R&D funding
4. Outcomes of State R&D Investments:
   A. Number of publications (total)
   B. Number of publications in referred journals
   C. Number and value of research proposals submitted
   D. Dollar value of research proposals submitted
   E. Number and value of research proposals submitted jointly with other Maine institutions
   F. Number and value of research proposals submitted jointly with non-Maine institutions
   G. Number and value of new federal research grants/contracts/subcontracts awarded (total)
   H. Number and value of new federal research grants/contracts/subcontracts awarded (EPSCOR only)
   I. Number and value of new federal research grants/contracts/subcontracts awarded (earmarked only)
   J. Number and value of new industrial research grants/contracts/subcontracts awarded (total)
   K. Number and value of new industrial research grants/contracts/subcontracts awarded (by Maine companies)
   L. Number and value of new foundation grants awarded
   M. Number of new companies formed on the basis of state supported R&D

(Continued)

BOX C-4 *Continued*

N. Number of jobs in these companies at spin-off
O. Number of disclosures made
P. Number of patents applied for
Q. Number of patents awarded
R. Number of copyrights obtained
S. Number of plant breeder's rights obtained
T. Number of licensing agreements completed (total)
U. Number of licensing agreements completed with Maine companies

Source: MSTF 2001

competitiveness. Two such programs that are federally funded but administered by universities are NSF's EPSCOR and the National Sea Grant Program.

The Experimental Program to Stimulate Competitive Research

EPSCOR is designed to improve the R&D competitiveness of states that have traditionally received smaller amounts of federal research and development funding—based on a per capita comparison. The program requires a commitment on the part of the states to improve the quality of science and engineering research and training at colleges and universities. Three key groups of metrics describe a state's science and technology environment: NSF support, total federal academic R&D contribution, and high technology activity (NSF 2002). Each group contains a number of metrics that can be compared across states and over time. Most of the metrics involve assessments in terms of people, programs, and dollars. The following are examples:

- Total number of NSF research-support awards per year.
- Academic R&D obligations by all federal agencies per year.
- Total number of graduate students in science and engineering.

Additional measures of the effectiveness of the programs include number of grant proposals submitted, number of grant proposals funded, quality of peer-reviewed research, professional contributions of students, publication and patent productivity, return on investment, and contribution to the state (for example, an improved environmental program) (NSF 2002).

National Sea Grant Program

The National Sea Grant Program, created in 1966, established a partnership between the National Oceanic and Atmospheric Administration and universities to encourage the development of sea-grant institutions for the purpose of engaging in research, education, outreach, and technology transfer in an effort to encourage stewardship of the nation's marine resources.

Performance benchmarks for evaluation have been developed to determine whether the goals and strategic plans of each sea-grant institution are being met. Programs are evaluated according to the following weighted criteria (NOAA, unpublished material, 1998):

- Effective and aggressive long-range planning (relative weight, 10%).
- Organizing and managing for success (relative weight, 20%).
- Connecting sea grant with users (relative weight, 20%).
- Producing significant results (relative weight, 50%).

Each sea-grant institution is given sets of recommended questions or established expected-performance benchmarks designed to gauge how well the program has met the goals established during strategic planning. Benchmarks typically include questions about the quality of the peer-review process, detailed information about the strategic planning process, measures to determine the quality of program management, ability of the program to develop private-sector matching funds, the number of published peer-reviewed papers in relation to the size of the research program, and questions to gauge the social, economic, and scientific contributions of program research (NOAA, unpublished material, 1998).

CONCLUSIONS

Federal research programs tend to focus more on the collection of product metrics than process metrics. Among federal research programs, there tends to be a presumption that peer review is the key process necessary to ensure a successful program. However, there tends to be relatively little discussion of who is responsible for conducting the peer-review evaluations. The committee considers that peer review is a necessary but not sufficient condition to ensure a successful program.

Evaluations at the state level are driven principally by economic considerations. There tends to be little targeting of specific research topics except in broad terms, such as nanotechnology. Many of the evaluations are based on surveys of participating institutions and data routinely collected at the state level, such as number of students enrolled in institutions of higher learning.

NSF's EPSCOR produces a level of standardization that allows comparison of R&D across states and across time. The standardization across time and place provides consistency, an important attribute of metrics.

REFERENCES

ARP/ATP (Advanced Research Program/Advanced Technology Program). 2002. Research Projects Performance Metrics. Texas Higher Education Coordinating Board, Austin, TX. [Online]. Available: http://www.arpatp.com/online/ [accessed June 12, 2002].

Burress, D., M. El-Hodiri, and V.K. Narayanan. 1992. An Evaluation Model to Determine the Return on Public Investment (ROPI) for the Kansas Technology Enterprise Corporation. Report No. 211. Institute for Public Policy and Business Research, The University of Kansas. November [Online]. Available: http://www.ukans.edu/cwis/units/ippbr/resrep/pdf/ M211.pdf [accessed Jan. 22, 2003].

MSTF (Maine Science and Technology Foundation). 2001. Evaluation of Maine's Public Investment in Research and Development. [Online]. Available: http: //www.mstf.org/evaluation/pdfjump.html. [accessed Jan. 29, 2003].

NIST (National Institute of Standards and Technology). 2002a. ATP Baseline Business Report. Optional Worksheet for Organizing Data. Advanced Technology Program, National Institute of Standards and Technology, Technology Administration, U.S. Department of Commerce.

NIST (National Institute of Standards and Technology). 2002b. ATP Anniversary Business Report. Optional Worksheet for Organizing Data. Advanced Technology Program, National Institute of Standards and Technology, Technology Administration, U.S. Department of Commerce.

NIST (National Institute of Standards and Technology). 2002c. ATP Close-Out Business Report. Optional Worksheet for Organizing Data. Advanced Technology Program, National Institute of Standards and Technology, Technology Administration, U.S. Department of Commerce.

NIST (National Institute of Standards and Technology). 2002d. ATP Post-Project Summary Business Report. Optional Worksheet for Organizing Data. Advanced Technology Program, National Institute of Standards and Technology, Technology Administration, U.S. Department of Commerce.

NSF (National Science Foundation). 2002. Experimental Program to Stimulate Competitive Research (EPSCoR). National Science Foundation. [Online]. Available: http://www.ehr.nsf.gov/epscor/statistics/ start.cfm/ [accessed Dec. 19, 2002].

Selden, R.W. 1998. Air Force Science and Technology Quality Review. Review Overview Document. Scientific Advisory Board Science and Technology. June.

USAFSAB/USAFRL (U.S. Air Force Scientific Advisory Board and U.S. Air Force Research Laboratory). 1998. Memorandum of Understanding for the Air Force Science and Technology Quality Review between U.S. Air Force Scientific Advisory Board, and U.S. Air Force Research Laboratory, Appendix III. Evaluation Criteria. August 1998.